liquid
scintillation
counting

liquid
scintillation
counting

K. D. NEAME
Department of Physiology,
University of Liverpool

C. A. HOMEWOOD
Department of Parasitology,
School of Tropical Medicine,
University of Liverpool

A HALSTED PRESS BOOK

JOHN WILEY & SONS
New York—Toronto

English edition first published in 1974 by Butterworth & Co
(Publishers) Ltd
88 Kingsway, London WC2B 6AB

Published in the U.S.A. and Canada by Halsted Press, a Division of
John Wiley & Sons, Inc.,
New York

Library of Congress Cataloging in Publication Data

Catalog card no. 74–11613

ISBN 0–470–63085–X

Text set in 10/12 pt IBM Press Roman, printed by photolithography,
and bound in Great Britain at The Pitman Press, Bath

Preface

A liquid scintillation counter is nowadays almost as common a piece of scientific equipment as a spectrophotometer. Yet in spite of this there is at the time of writing no single book which describes the principles and elementary practice of its operation. Most beginners have to learn what they can from manufacturers' instruction manuals, from published papers and booklets, and from the advice of colleagues.

This book developed in fact as a result of arguments and discussions between the authors over points, many of them trivial, which could usually only be resolved either by experiment or by hours of often fruitless searching through published reports. It has been written in the hope of saving others from similar time-wasting efforts. We do not claim that the book is exhaustive; on the contrary, we have deliberately simplified the wealth of often contradictory information, and in the experimental sections have given what we believe to be a few useful and well-tested methods instead of trying to list them all. We have checked on as many practical details as possible in our own laboratories, but in spite of this we have felt unable to be dogmatic over many points. Not only do liquid scintillation counters vary in their characteristics, and therefore in the uncorrected data which they produce, but it is often impossible to predict the effects of many other sources of variation except in a general way. We have therefore tried to suggest possible sources of error; the precise effects of the many variables on an individual experiment must be evaluated by the experimenter.

Where possible, we have given references for further reading and to more specialised techniques, but in many places no references to specific statements have been given, since much of the information in the text is derived from the experience of our colleagues and ourselves and from other unpublished sources. We are therefore indebted to many people for information, discussions and advice. We are particularly grateful to the following: to Dr. G. A. Buckley of the Department of Pharmacology, University of Liverpool, for many helpful discussions and for reading the manuscript; to Mr. I. Birchall, Radiation Protection Officer of the University of Liverpool, who also offered much helpful advice on the manuscript; to Mr. K. Wilkinson of the Department of Physiology, University of Liverpool, to Mr. B. Lumb of Nuclear-Chicago Ltd. (G. D. Searle & Co.), High Wycombe, and to Dr. J. C. Turner of the Radiochemical Centre, Amersham, for helpful discussions and information; to Mrs. V. C. Baggaley, Miss E. M. Atkinson and Mrs. P. A. Brownbill of the Department of Parasitology, Liverpool School of Tropical Medicine, who patiently and competently carried out many of the experiments and also criticised the manuscript; to Packard Instrument Ltd., New England Nuclear, Nuclear Enterprises Ltd. and Nuclear-Chicago Ltd. for kind gifts of materials; and to the heads of our respective departments, Prof. R. A. Gregory (who suggested the writing of this book) and Prof. W. Peters, for their enthusiastic support. Any mistakes are, of course, entirely the responsibility of the authors, who would be pleased to learn of them, together with any further comments or suggestions.

K.D.N.
C.A.H.

ACKNOWLEDGEMENTS

We would like to thank the following for kindly allowing us to include in some of the Tables data from other published work: The Radio-chemical Centre, Amersham (Tables 2.1, 2.7 and 2.8); ICN Pharmaceuticals (U.K.) Ltd., Tracerlab Instruments Division (Table 2.2); Dr. J. M. A. Lenihan and Pitman Publishing (Table 2.2); Dr. J. Mantel (Table 2.5); Pergamon Press (Tables 2.3 and 2.5); and Drs. C. H. Wang and D. L. Willis, and Prentice-Hall, Inc (Tables 2.4 and 3.1).

Contents

Technical Note

The experimental data in the illustrations in this book (except for Figs. 7.6 and 7.7) were obtained in the authors' laboratories on one or other of two Nuclear-Chicago Unilux II Liquid Scintillation Spectrometers (G. D. Searle Ltd., High Wycombe, England). The counting chamber of this instrument is maintained at about 5°C. There are two photomultiplier tubes whose voltage is fixed by the manufacturer. Amplification is linear, and coincidence counting and pulse summation operate. Attenuation (gain) in each of three analysing channels is effected by means of a coarse 'click-stop' control, with a factor of two between each position, in conjunction with a ten-turn helipot control providing continuous fine attenuation which overlaps each of the coarse settings. The discriminators cover the range $0 \cdot 1 - 9 \cdot 9V$ in discrete steps of $0 \cdot 1$ V; the upper discriminator may be made inoperative. The external standard is ^{133}Ba.

It is important to realise that it may not be possible to reproduce exactly on other instruments the data shown in the diagrams in this book, although qualitatively similar results should be obtained. Liquid scintillation counters even of the same design may differ greatly, each instrument having its own personal characteristics. For example, on one of our instruments the maximum counting efficiencies for ^{14}C and ^{3}H were found to be about 95% and 65% respectively, and on the other about 90% and 55%; these differences were even more marked with quenched samples.

CHAPTER ONE

Introduction

One of the most fruitful and versatile ways of investigating a wide range of physical and chemical phenomena is to use a radioactive form of the material under investigation. A small quantity of this radioactive form, the *tracer*, is usually mixed with a much larger amount of its non-radioactive counterpart, and the behaviour of the two in the experimental system is then assumed to be identical. Since the tracer can be relatively easily detected by its radioactivity, it can provide information about the behaviour of the non-radioactive form which it may not be possible to obtain by more conventional methods. For example, by the use of tracers, individual atoms of a molecule can be followed through a series of reactions. Furthermore, molecules which differ only in the time at which they became incorporated into a population may be distinguished from each other.

For the use of tracers to be most effective, a sensitive method of detecting radioactivity is required. One such method is liquid scintillation counting, which allows very small amounts of radioactivity to be measured with accuracy. The method does no more, however, than measure radioactivity, and gives no indication of the chemical form of a radioactive material. Conventional chemical and physical techniques are thus still essential for the separation and identification of this material.

It is all too common for the newcomer to liquid scintillation counting to believe that the measurement of radioactivity is somehow magically different from other methods of measurement, and that the mere use of radioactive tracers is itself enough to guarantee meaningful results. In practice, of course, the reliability of experimental results depends entirely on the accuracy of their measurement, and with a liquid scintillation counter, as with any other measuring instrument, this accuracy can only be achieved by an understanding of the basic principles of its operation and by the comprehensive use of controls and standards.

Because radioactive decay is known to be almost entirely unaffected by any external conditions ordinarily encountered, it is often assumed that its measurement is similarly unaffected and this leads to a belief that no special care is needed in judging raw data. This belief is encouraged by the mystique and special vocabulary which have grown up around scintillation counting, so that hard-learned and well-tried methods of approach may be discarded in the belief that they do not apply to the assay of radioactivity. Nothing could be further from the truth. Liquid scintillation counting is merely a sensitive method of measurement, which not only requires all the usual basic precautions associated with any such method, but in addition requires special care, because of certain problems peculiar to the technique.

Much of the specialised vocabulary used in liquid scintillation counting consists of new ways of expressing well-established concepts. This can be seen from Table 1.1, in which terms used in liquid scintillation counting are compared with terms having equivalent meanings in spectrophotometry.

TABLE 1.1. Comparable concepts in liquid scintillation counting and spectrophotometry

Liquid scintillation counting	Spectrophotometry
Background count rate	Blank value
Count rate	Optical density
Effect of scintillator concentration on count rate	Effect of reagent concentration on colour development
Linear relationship between count rate and specific activity	Linear relationship between optical density and concentration
Optimum fluid volume in counting vial	Optimum fluid volume in cuvette
Optimum settings of discriminators	Optimum wavelength
Primary solvent and scintillator	Colour reagent
Pulse voltage spectrum	Colour absorption spectrum
Quenching	Interference with colour development
Specific activity	Concentration
Stability of count rate	Stability of colour

Any description of the liquid scintillation counter and its use must only be taken as a general guide. Instruments vary in design and in performance, so that uncorrected experimental data may vary considerably from one instrument to another. We have even found this with two supposedly identical instruments which we have used for most of the illustrations in this book (*see* Technical Note, p. viii).

Most of the descriptions and illustrations which follow are based on one type of liquid scintillation counter. In this type, amplification is linear, in contrast to the alternative type in which amplification is logarithmic (p. 35). A text which described the use of each separately would be unwieldy and unnecessarily long. The operation of an instrument with linear amplification is the more complicated of the two, and the following descriptions can be used for an instrument with logarithmic amplification by the omission of certain details which are pointed out where applicable.

CHAPTER TWO

Radioactivity

Before radioactive material can be used to maximum advantage and be handled with safety, it is necessary to know something of the nature of radioactivity itself and of the particles and rays emitted.

Radioactive atoms are characterised by the instability of their nuclei. The nucleus of an atom consists of a tightly bound group of protons and neutrons in continuous motion. Every atomic species can be described in terms of the constituents of its nuclei, and it is then referred to as a *nuclide*. Nuclides can be divided into groups on the basis of the number of protons in the nucleus, and each group then has the chemical properties of a single element; the members of each group are known as the *isotopes* of that element. Thus the term nuclide can be used when referring to *any* atomic species, but the term isotope is restricted to the various forms of a single element.

All the isotopes of an element have the same number of protons but differ in the number of neutrons in their nuclei. As a result, the physical properties of the various isotopes of a single element are not absolutely identical[9], because nuclear mass, nuclear volume and density of nuclear charge, upon which these properties depend, are different in each. For example, the common, stable, form of hydrogen, whose nucleus contains one proton but no neutrons, has only about one-third of the mass of the radioactive form (tritium) whose nucleus contains one proton and two neutrons (Table 2.1). (Protons and neutrons have very nearly the same mass.) The common form of hydrogen may consequently be expected to diffuse nearly twice as fast in the gaseous state as the radioactive form, since the rate of gaseous diffusion is inversely proportional

TABLE 2.1. Properties of the isotopes of hydrogen and carbon[108]
(Stable isotopes in heavy type)

		Hydrogen			Carbon					
		^{1}H	^{2}H*	^{3}H**ϕ	^{10}C	^{11}C	^{12}C	^{13}C	^{14}Cϕ	^{15}C
Symbol										
Protons (atomic number, Z)		1	1	1	6	6	6	6	6	6
Neutrons		0	1	2	4	5	6	7	8	9
Total particles (mass number, A)		1	2	3	10	11	12	13	14	15
Abundance in nature (%)		99·98	0·02	–	–	–	98·9	1·1	–	–
Half-life		stable	stable	12·3 yr	19 sec	20 min	stable	stable	5760 yr	2·3 sec
Emission		–	–	β^-	$\beta^+;\gamma$	β^+	–	–	β^-	$\beta^-;\gamma$
(Maximum) energy of emission (MeV)	β^-	–	–	0·018	1·9	0·96	–	–	0·158	4·3, 9·8
	γ	–	–	–	0·72, 1·03	–	–	–	–	5·3
Daughter nuclide		–	–	^{3}He	^{10}B	^{11}B	–	–	^{14}N	^{15}N

* Deuterium
** Tritium
ϕ Used as tracers

to the square root of the molecular weight[36]. Differences such as this can be used to separate the isotopes of an element, although in some cases radioactive isotopes are prepared by methods which do not require such separation.

Nuclides may be stable or unstable. An unstable nuclide (*radionuclide** or, when referring to an isotope, *radioisotope**) spontaneously changes or *decays*, at some unpredictable time, to a more stable form. This decay is observed by the detection of particles or electromagnetic radiation emitted by the nuclide. The decay with its associated emissions is known as *radioactivity*, and the resulting nuclide is called a *daughter* (*nuclide*). The daughter may itself also decay, and a series of radioactive nuclides may be produced before a stable form is finally reached.

The stability of a nucleus depends on the relative proportions of protons and neutrons; broadly speaking, the more these proportions diverge from the stable values, the greater the instability, and the sooner the decay (*see*, for example, the isotopes of carbon in Table 2.1).

Most of the nuclides found in nature are stable, since unstable nuclides eventually decay to a stable form, although collisions between cosmic rays and stable nuclei maintain a continuous natural supply of unstable nuclides. Radionuclides for research, however, are usually synthesised from stable nuclei in atomic reactors or, in certain cases, in nuclear accelerators.

The notation which we shall use for designating a particular nuclide is that recommended by the International Union of Pure and Applied Chemistry[53], and consists of the usual chemical symbol for the element preceded by a superscript indicating its mass number, that is, the number of neutrons plus the number of protons. Thus the common, stable, form of carbon, which has six neutrons and six protons, is written ^{12}C, while its long-lived radioisotope, with eight neutrons and six protons, is written ^{14}C (*see*, for example, Table 2.1).

A radioisotope can be incorporated into a molecule in place of a stable isotope. Its position in the molecule may be denoted in a number of ways, but the practice commonly adopted by many chemical and biochemical authorities is to place the symbol for the isotope "in square brackets directly attached to the front of the chemical name. The isotopic prefix precedes that part of the name to which it refers, as in sodium $[^{14}C]$ formate . . . . The positions of isotopic labelling are indicated by . . . prefixes placed within the square brackets and before the symbol of the element concerned" (*Radiochemicals Catalogue* of

---

* In this chapter we shall use each of these two terms strictly in accordance with its specific meaning. In subsequent chapters we shall follow common usage and refer to a radioactive element as a radioisotope, since it will then usually be discussed in unspoken relationship with the corresponding stable isotope.

the Radiochemical Centre, Amersham, 1973/4), for example, $n-[1-^{14}C]$ hexadecane.

TYPES OF RADIOACTIVE DECAY

There are a limited number of ways in which an unstable nucleus may decay.

Some radioactive elements, particularly those of heavier mass, may emit a tightly bound complex of two protons and two neutrons, known as an α-particle. (A few very heavy nuclei split into two distinct parts of unequal mass.) Others, particularly those of lighter mass, may emit either an electron or the corresponding positively charged particle called a *positron*. Electrons and positrons emitted in radioactive decay are referred to as β- (β^-- and β^+- respectively) *particles*.

A less common method of moving towards stability is the absorption of an orbital electron into the nucleus; the rearrangement of the remaining electrons results in the emission of an X-ray.

A nucleus may also lose surplus energy in the form of electromagnetic radiation of very short wavelength (γ-rays), usually as a secondary process accompanying one of the other decay processes just mentioned.

These emissions will now be discussed briefly, since it is by their detection that radionuclides are measured; their properties are summarised in Table 2.2.

α-particles

These are indistinguishable from nuclei of the stable isotope of helium, ^{4}He. Since they contain two protons they have a double positive charge. The nucleus which remains after the emission of an α-particle has two protons and two neutrons less than the parent; it is two places lower in the periodic table and four units less in mass number. For example, the radioactive gas radon, ^{222}Rn, is derived in this way from the decay of the common form of radium, ^{226}Ra.

α-particles are readily stopped by interaction with matter, causing intense local ionisation.

β-particles

The emission of a β^--particle results in the gain of a positive charge by the nucleus, and this is associated with the conversion of a neutron (n)

TABLE 2.2. Some properties of nuclear emissions[61, 62]

Emission	Nature	Charge	Example of source	Energy, typical values (MeV)	Range approx. (mm)		
					Air	Water	Sodium Iodide*
α-particle	Helium nucleus	++		4–9	30–90	0·025–0·09	–
			^{210}Po	5·3	38	0·03	0·04
β⁻-particle (negatron)	Electron	–		0·02–5	4–11 000	0·009–15	–
			^{3}H	0·018 (max)	4	0·007	0·002
			^{14}C	0·16 (max)	280	0·3	0·08
			^{32}P	1·7 (max)	6000	8	2
β⁺-particle (positron)	Positive electron	+		0·3–4	Mutual annihilation with electron in 10^{-9} sec		
γ-ray	Electromagnetic radiation	Nil		0·01–5	(15–50) × 10^{3}***	50–200**	–
			^{125}I	0·03	–	60**	1**
			^{51}Cr	0·3	–	170**	42**

* Commonly used as detector for γ-rays in crystal scintillation counters
** Distance for reduction of intensity to 1/10 of original value; in lead, values are about 8–60 mm for γ-rays whose energies are 0·4–1·3 MeV[90]

to a proton (p), that is,

$$n \rightarrow p + \beta^-$$

In this way, the radioactive ^{14}C (six protons, eight neutrons) becomes the stable ^{14}N (seven protons, seven neutrons) (Table 2.1), thus:

$$^{14}C \ (6p + 8n) \rightarrow {}^{14}N \ (7p + 7n) + \beta^-$$

The ejection of a β^+–particle, on the other hand, results in the loss of a positive charge by the nucleus, and is associated with the conversion of a proton to a neutron, that is,

$$p \rightarrow n + \beta^+$$

For example, the radioactive ^{11}C (six protons, five neutrons) becomes the stable ^{11}B (five protons, six neutrons) in this way (Table 2.1):

$$^{11}C \ (6p + 5n) \rightarrow {}^{11}B \ (5p + 6n) + \beta^+$$

The type of β–particle emitted by a particular nucleus depends mainly on the proportions of protons and neutrons. When protons are in relative excess, they are more likely to be converted to neutrons, with the loss of a β^+–particle, whereas when there is a relative excess of neutrons, the conversion of neutrons to protons with the loss of a β^-–particle is more probable (compare ^{10}C and ^{11}C with ^{14}C and ^{15}C in Table 2.1).

From the practical point of view there are two major differences associated with these particles. Firstly, a β^+–particle is rapidly annihilated by collision with an orbital electron; the opposite charges annul one another and the total mass is converted to energy, which appears as γ–rays of a particular wavelength[36]. A β^-–particle, on the other hand, survives, since it is unlikely to encounter a β^+–particle (these have a life of about 10^{-9} seconds). Secondly, most of the radionuclides from which β^+–particles are derived are produced in accelerators rather than atomic reactors, and are consequently less freely available as tracers.

β^-–particles are stopped fairly readily by interaction with matter; those of lower energy are unable to penetrate the wall of a glass vessel of normal thickness.

γ–rays

These are emitted from the nucleus in many radioactive decay processes and differ from other forms of electromagnetic radiation only in the extreme shortness of their wavelengths (about $0.001-0.1 \ m\mu$).

X—rays

These are produced *inter alia* during the type of decay known as *electron capture* (E.C.), in which an orbital electron is captured by an unstable nucleus, causing the transformation of a proton to a neutron, thus:

$$p + e^- \rightarrow n$$

(When, as often happens, the electron originates from the K—shell, the term *K—capture* may be used.) The resulting rearrangement of the remaining orbital electrons causes the emission of an X—ray. An example of this is the conversion of the radioactive ^{55}Fe to the stable ^{55}Mn.

The only difference between emitted X—rays and γ—rays of similar wavelength lies in the nature of their production, the first being derived from the rearrangement of orbital electrons, the second from forces within the nucleus. X—rays generally have lower energies (longer wavelengths) than γ—rays, so that shielding from them is easier.

ENERGY OF RADIOACTIVE EMISSIONS

The particles and electromagnetic radiation emitted by a decaying nucleus carry a certain amount of energy, which is usually measured in units of *million electron volts* (MeV) or *thousand electron volts* (keV), where one electron volt is equivalent to the energy which would be acquired by an electron moving through a potential of one volt. α— and β—particles carry kinetic energy, that is, energy related to the velocity of the particle.

An α—particle or a γ—ray emitted from a given radionuclide has an energy with one of a limited number of sharply defined values that are characteristic of that radionuclide. In contrast, β—particles emitted by a specified radionuclide have a continuous range of energies. This is because the energy is shared in an unpredictable way between the β—particle and a neutrino which is expelled at the same time. (The neutrino is an elusive particle with velocity and spin, but with extremely small mass. It is not normally detectable; in fact, it has been calculated[31] that on average it would pass through 50 light years of lead before being stopped.) However, the sum of energies of β—particle and neutrino is believed to be constant for any particular decay pattern of a radionuclide; it follows that in the limit, when a neutrino is not emitted, all the energy is carried by the β—particle. β—particles thus have a range of energies from zero up to a maximum whose value is constant for a particular decay pattern. This maximum value (*max.* MeV) is commonly quoted as the β—particle energy (Table 2.3).

Radio-isotope	Common stable isotope	Emission	Energy (or maximum energy) of major emissions* MeV	Half-life			Maximum permissible body burden ϕ μCi	Organ in which concentrated
				Physical†	Biological ϕ	Effective ϕ (in body)		
^{3}H (tritium)	^{1}H	β^-	0·018	12·3 yr	12 dy	12 dy	1000 (as ^{3}H$_2$O)	Whole body
^{14}C	^{12}C	β^-	0·159	5760 yr	10–40 dy	10–40 dy	300 (as ^{14}CO$_2$)	Whole body
^{22}Na	^{23}Na	β^+ γ	{0·54, 0·51, 1·28}	2·6 yr	11 dy	11 dy	10	Whole body
^{32}P	^{31}P	β^-	1·71	14·3 dy	3·3 yr	14 dy	6	Bone
^{35}S	^{32}S	β^-	0·167	87·2 dy	1·7 yr	76 dy	90	Testis
^{36}Cl	^{35}Cl, ^{37}Cl	β^-	0·714	300000 yr	29 dy	29 dy	80	Whole body
^{45}Ca	^{40}Ca	β^-	0·254	165 dy	49 yr	162 dy	30	Bone
^{51}Cr	^{52}Cr	X-ray γ	{0·005, 0·323}	27·8 dy	1·7 yr	27 dy	800	Whole body
^{55}Fe	^{56}Fe	X-ray	0·006	2·7 yr	1·5–4·6 yr	1·1–2·2 yr	1000	Spleen
^{59}Fe	^{56}Fe	β^- γ	{0·27, 0·46; 1·10, 1·29}	45 dy	1·5–4·6 yr	43 dy	20	Spleen
^{63}Ni	^{58}Ni, ^{60}Ni	β^-	0·067	120 yr	1·4–1·8 yr	1·4–1·8 yr	200	Bone
^{89}Sr	^{88}Sr	β^-	1·46	51 dy	35–49 yr	50 dy	4	Bone
^{90}Sr	^{88}Sr	β^- (including daughter ^{90}Y) (trace γ)	0·54, 2·27	28 yr	35–49 yr	18 yr	2	Bone
^{125}I	^{127}I	X-ray γ	{0·027, 0·035}	60 dy	138 dy	41 dy	10	Thyroid
^{131}I	^{127}I	β^- γ	{0·61, 0·36}	8·0 dy	138 dy	7·6 dy	0·7	Thyroid
^{210}Po	—	α γ	{5·3, 0·8}	138 dy	70 dy	46 dy	0·03	Spleen, kidney

* Percentage of each may be found in the *Radiochemical Manual*[108]

† 'Half-life' as commonly used usually refers to the physical half-life

ϕ The biological half-life, the effective half-life and the maximum permissible body burden depend on the nature of the compound into which the radionuclide is incorporated, and may vary considerably with certain radionuclides, particularly ^{2}H, ^{14}C and ^{35}S

With some radionuclides every nucleus decays in the same basic way. With others, however, there may be a number of possible patterns of decay, and for each such radionuclide the proportion of nuclei which decay in any particular way is constant (Table 2.4). Thus a particle may be emitted either alone or with one or more γ—rays. However, the total

TABLE 2.4. Alternative decay routes
Energies of alternative decay routes of an α—particle emitter and a β—particle emitter, and the proportions in which they are found[104]. Each nucleus emits either an α— or a β—particle, which may be followed by one or more γ—rays. These may have various energies, but provided that the final product is the same, the total energy is constant

| | *Energy of emissions* MeV *(α—particles, γ—rays) or* max. MeV *(β—particles)* | | | | | | | |
	²¹⁹Rn *(α—emitter)*			¹³¹I *(β⁻—emitter)*				
α—particles	6·81	6·55	6·42	—	—	—	—	—
β—particles	—	—	—	0·61	0·61	0·33	0·25	0·81
γ—rays {	—	0·27	0·40	0·36	0·28	0·64	0·72	0·16
{	—	—	—	—	0·08	—	—	—
	6·81	6·82	6·82	0·97	0·97	0·97	0·97	0·97
Proportion of nuclei decaying by route in question	82%	13%	5%	87%		9%	3%	1%

energy emitted by each nucleus of a given radionuclide decaying to a given product nucleus is the same, and when there is more than one type of emission this energy is distributed between them. In the case of a β—particle emitter the total energy of emission is the sum of the γ— ray energy and the *maximum* energy of the β—particles.

Differences in the energies of emissions may be used to analyse a mixture of radionuclides. It may be relatively simple to distinguish between the α—particles from different radionuclides because their energies are discrete; the same is true of γ—rays. However, the resolution of data from a mixture of β—emitting radionuclides is more difficult because of the continuous nature of the energy spectra of β—particles. Only if the spectra of energies of two β—emitters cover greatly different energy ranges, as for example with ¹⁴C and ³H (Table 2.5), can they be readily distinguished. Even in such cases, however, both radionuclides emit some β—particles of low energy, so that distinction between them can only be made over the upper range of energies. For resolution to be satisfactory, therefore, there must be adequate differences in the energy of the majority of the particles, that is, differences in the mean energy. The mean and maximum energies of β—particles emitted by some radio-nuclides are shown in Table 2.5.

TABLE 2.5. Energy of β–particles emitted by some commonly used radionuclides[67] (Owing to the difference in source of reference, the β–particle energies shown do not all correspond exactly to those given in Table 2.3.)

Radionuclide	β–particle energy (MeV)		Relative mean energy (^3H taken as unity)
	Mean	Maximum	
^3H	0·0057	0·019	1
^{35}S	0·049	0·167	8·6
^{14}C	0·049	0·156	8·6
^{45}Ca	0·077	0·254	13·5
$^{22}Na*$	0·215	0·545	37·7
^{36}Cl	0·321	0·714	56
^{32}P	0·693	1·71	122

* Positron emission

PENETRATION OF MATTER BY RADIOACTIVE EMISSIONS

Radioactive emissions interact with atoms as they pass through matter, transferring some of their energy in the process. This is important when considering the ways in which radioactivity may be detected, the interference which an outside source of radiation may exert on the measurement of radioactivity, and the radiation to which a person may be exposed.

The extent of the penetration of matter depends on the type of radiation, on its energy and on the nature of the material through which it passes (Table 2.2). In general, γ–rays are more penetrating than β–particles, and these in turn are more penetrating than α–particles. The greater the energy of a given type of emission, the greater is its range, but this is also affected by the atomic number, the mass number or the density of the material, depending on the type of emission.

γ–rays and X–rays

When passing through uniform matter the intensity of γ–rays and X–rays in general decreases logarithmically with distance. When radiation of this type passes through air, the loss is relatively small; when derived from a point source, its intensity roughly follows the inverse square law of electromagnetic radiation *in vacuo*, that is, the energy passing across unit area in unit time decreases inversely as the square of the distance from the source. Theoretically the intensity becomes zero only at infinite distance from the source, but there is in practice a distance at which the radiation is no longer detectable.

Low energy γ—rays (less than 1 MeV) are most effectively absorbed by materials of high atomic number and high density, whereas the atomic number is of less importance with γ—rays of high energy. Lead, which has a high atomic number and high density, is commonly used to shield against γ—rays; concrete is also an effective shield.

The terms 'hard' and 'soft' are often used as a loose description of the ability of γ—rays or X—rays to penetrate matter, 'hard' rays having a higher energy and penetrating more readily than 'soft' rays.

A common unit of reference for the penetration of solid material by γ—rays is the *half-thickness*, which is the thickness of material which reduces the intensity of the radiation by half. A more useful unit from the point of view of shielding is the distance in which the intensity is reduced to one-tenth of its original value (Table 2.2). For example, the intensity of γ—rays emitted by ^{60}Co (1·17 and 1·33 MeV) is reduced to one-half of its original value by about 12 mm of lead and to one-tenth by about 40 mm.

α— and β—particles

The penetrating power of α— and β—particles is much less than that of most γ—rays (Table 2.2) and their range has a definite limit, when virtually all their energy has been lost as a result of collisions with atoms. The range of most α—particles is only about 3—4 cm in air at ordinary temperature and pressure, and for any given α—particle energy it is fairly sharply defined. It is much less than that of β^-—particles of the same energy, which have a range of up to a few metres in air. However, the distances travelled by the β^-—particles emitted from a given radio-nuclide vary; this is mainly owing to the spread of energies but also partly to scattering in the material through which they pass. β^-—particles also generate weak X—rays (Bremsstrahlung) when they pass through matter and these may be more penetrating than the particles themselves.

The shielding of β—particles with energies up to about 2 MeV and of α—particles is not difficult, since they are in most cases effectively stopped by materials such as glass.

RATE OF DECAY AND HALF-LIFE

Radioactive decay is random; there is no way of telling when any particular nucleus will decay. All that can be said is that with a large enough group of nuclei there is a high probability that a given proportion of them will decay in a specified time. (The larger the group, the more accurate will be the prediction.) This proportion or fraction, known as the *decay constant*, is peculiar to each radionuclide. The amount of a radionuclide present decreases exponentially: although the proportion of nuclei decaying in any series of equal time intervals is constant, the number of nuclei remaining falls progressively. For any given radionuclide, the time in which half the nuclei have decayed has a constant value; this is known as the *half-life* of that radionuclide (*see* Table 2.3).

Correction for the loss due to decay is not important in those tracer studies in which a slowly decaying radionuclide such as ^{14}C is used, since the change in amount between preparation and final measurement is negligible. With many radionuclides, however, decay may be fast enough to produce significant changes during storage or even during the course of an experiment. For example, in 2 years about one-tenth of the amount of ^{3}H initially present has decayed; the same fraction of ^{24}Na is lost in about $2\frac{1}{2}$ hours. In such cases, it may be necessary to take into account the change in amount of radioactive material present.

DETERMINATION OF LOSS DUE TO DECAY

The loss of radioactive material due to decay may be determined either graphically or by calculation, as shown below.

Graphical method

The amount of radioactivity present is plotted on either a linear or a logarithmic scale against the elapsed time; an example, using ^{3}H, is shown in Fig. 2.1. A semilogarithmic plot is preferable, for then the exponential decay of a single radionuclide appears as a straight line which may be drawn simply by plotting through two points, one of which represents the amount of radionuclide present at time zero, while the other represents half this amount at a time equal to the half-life of the radionuclide. The amount of radionuclide remaining after any chosen time can then be determined from the graph.

Calculation method

The rate of decay may be expressed by the differential equation

$$dN/dt = -\lambda N \qquad (2.1)$$

where N is the number of radioactive atoms at time t and λ is the decay constant. On integration this appears as

$$N/N_0 = \exp(-\lambda t) \tag{2.2}$$

where N_0 is the number of radioactive atoms originally present, and N/N_0 is the fraction remaining after time t.

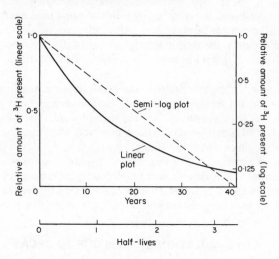

Fig. 2.1. *Graphical representation of radioactive decay. The curved line shows the decay of ^3H on a linear plot* (left-hand scale) *and the straight line the decay on a semilogarithmic plot* (right-hand scale)

It is more useful to express this in terms of the half-life, whose value is more readily available than that of the decay constant. If t is made equal to the half-life, $t_{\frac{1}{2}}$, then by definition $N/N_0 = 0.5$, and Equation 2.2 appears in this special case as

$$0.5 = \exp\left(-\lambda t_{\frac{1}{2}}\right) \tag{2.3}$$

from which

$$\log_e 0.5 = -0.693 = -\lambda t_{\frac{1}{2}} \tag{2.4}$$

so that

$$\lambda = 0.693/t_{\frac{1}{2}}. \tag{2.5}$$

Equation 2.2 may now be expressed as

$$N/N_0 = \exp\left(-0.693t/t_{\frac{1}{2}}\right) \tag{2.6}$$

or

$$\log_e (N/N_0) = -0.693t/t_{\frac{1}{2}} \tag{2.7}$$

or, since $10^1 = e^{2.303}$,

$$\log_{10} (N/N_0) = -0.693t/2.303t_{\frac{1}{2}}. \tag{2.8}$$

The fraction of radioactive material remaining after time t may then be determined from any of these last three equations.

Table 2.6 is a decay table for ^{3}H and shows the fraction remaining after intervals up to 6 years.

TABLE 2.6. Decay table for ^{3}H
Values represent the fraction remaining after the time indicated, calculated from a decay constant of 0·0047/month

Months	Years					
	0	1	2	3	4	5
0	1·000	·945	·893	·844	·798	·754
1	·995	·941	·889	·841	·794	·751
2	·991	·936	·885	·837	·791	·747
3	·986	·932	·881	·833	·787	·744
4	·981	·928	·877	·829	·783	·740
5	·977	·923	·873	·825	·780	·737
6	·972	·919	·869	·821	·776	·733
7	·968	·915	·865	·817	·772	·730
8	·963	·910	·860	·813	·769	·727
9	·959	·906	·856	·810	·765	·723
10	·954	·902	·852	·806	·762	·720
11	·950	·898	·848	·802	·758	·716

UNITS OF RADIOACTIVITY

The decay of a single nucleus, associated with the emissions already described, is commonly known as a *disintegration*. (The term should not be confused with nuclear fission, a particular type of disintegration in which the nucleus of an unstable heavy element breaks into two unequal parts.) The rate of decay of a radioactive material is known as its *activity*, which is defined as the number of nuclear transformations which occur in a specimen of radioactive material in unit time[73].

The basic unit of activity is the *curie* (Ci), defined as 3·700 x 10^{10} *disintegrations per second* (dps). (This is the rate of disintegration of 1 g of radium, although the curie was originally defined as the mass of its derivative radon in equilibrium with this amount of radium.) The curie is a large unit for radioactive tracer work, and smaller units are more useful; these are the *millicurie* (mCi), 1 Ci x 10^{-3}, and the *microcurie* (μCi), 1 Ci x 10^{-6}. (Also sometimes used are the nanocurie, 1 Ci x 10^{-9}, and the picocurie, 1 Ci x 10^{-12}.) Activity is also expressed as *disintegrations per minute* (dpm); then 1 mCi = 2·220 x 10^9 dpm and 1 μCi = 2·220 x 10^6 dpm.

Specific activity

It is often necessary to state the amount of radioactive material in terms similar to concentration, that is, per unit quantity of material, and this is called the *specific activity*. There are no universally accepted units for this, but it is usually expressed in terms of the rate of disintegration either (1) per unit quantity of the element in question (radioactive plus non-radioactive isotopes), that is, dpm/mole, Ci/mole, etc., or (2) per unit mass volume of material containing the radionuclide, that is, Ci/ml, Ci/g, etc. Sometimes specific activity is expressed in terms of count rate per unit quantity of material. In view of this ambiguity, the units must always be stated.

The highest possible specific activity of a compound containing a given radioactive *label* (that is, with a specified atom or atoms replaced by a radioisotope) is obtained when every molecule contains the radio-isotope, that is, when there are no unlabelled molecules present. When each molecule contains, for example, one radioactive atom (or label) the maximum specific activity may be calculated in the following way. One mole of the compound contains 6.023×10^{23} radioactive atoms (Avogadro's number). The fraction of these atoms which decay in unit time is given by the decay constant, λ, and the number of atoms which disintegrate in one mole in unit time is therefore $\lambda \times 6.023 \times 10^{23}$. This is the maximum specific activity of that compound. For example, if every molecule of a compound were labelled with one atom of ^{14}C (decay constant 0.00012/yr) the rate of decay would be

$$(6.023 \times 10^{23}) \times 0.00012 \text{ disintegrations per mole per year} =$$

$$\frac{(6.023 \times 10^{23}) \times 0.00012}{365 \times 24 \times 60} \text{ disintegrations per mole per minute} =$$

$$\frac{(6.023 \times 10^{23}) \times 0.00012}{365 \times 24 \times 60 \times (2.22 \times 10^{12})} \text{ Ci/mole} =$$

62 Ci/mole.

A similar calculation shows that the specific activity of a compound in which each molecule contains one atom of ^3H is 29500 Ci/mole. This difference between a $^3H-$ and a $^{14}C-$labelled compound is a reflection of the different rates of decay. Thus a much greater propor-tion of the atoms of ^3H, which has a relatively short half-life, are decaying in any unit of time than are those of ^{14}C with its much longer half-life.

TABLE 2.7. Maximum specific activity and cost of some commercially available radiochemicals
Note that the maximum specific activity in the units quoted is inversely proportional to the half-life of the radionuclide incorporated.
Commercial data reproduced from the *Radiochemicals Catalogue* (1973/4), The Radiochemical Centre, Amersham, with permission

Radiochemical	Cost				Maximum specific activity Ci/mole		Approximate half-life
	50 μCi	250 μCi	1 mCi	5 mCi	Commercially available	Theoretical	
D–[1–^{14}C] glucose	£12	£45	—	—	>40		
D–[2–^{14}C] glucose	£18	£75	—	—	20–35	62	5 760 yr
D–[6–^{14}C] glucose	£14	£60	—	—	45		
D–[1–^{3}H] glucose	—	—	£12	£30	3 000		
D–[2–^{3}H] glucose	—	£7	£15	£50	100–500	29 500	12 yr
D–[6–^{3}H] glucose	—	£7	£12	£30	2 000–10 000		
L–[^{35}S] methionine	—	—	£35	£95	5 000–25 000	1 500 000	0·24 yr

It follows that radioactive chemicals (*radiochemicals*) which contain a radionuclide with a relatively rapid rate of decay (such as ^{32}P, ^{35}S or ^{3}H) should be, and to a large extent are, available at higher specific activity than are those containing a longer-lived radionuclide (such as ^{14}C or ^{36}Cl), and at a lower cost per curie, assuming similar methods of manufacture (Table 2.7). This advantage may be offset by a much faster loss of activity and a more rapid chemical deterioration during storage (p. 25), and in the case of ^{3}H by a reduced accuracy of measurement owing to low β—particle energy (*see* Chapter 4).

Most commercially available radioactive compounds have a specific activity below the theoretical maximum (Table 2.7). This may be because non-radioactive compound has been added to reduce decomposition (p. 26) or it may be due to difficulty in separating radioactive from stable isotope during preparation. Non-radioactive material having the same or similar properties to the radioactive material and with which it is mixed is called a *carrier*. In this case it is added to protect the radioisotope or to make it easier to handle. Radiochemicals which contain no carrier (*carrier-free*) are also available in some cases, although the term 'carrier-free' is sometimes loosely used to refer to material which merely has high specific activity.

In tracer investigations, 'carrier' usually refers to the non-radioactive compound whose properties are being investigated by means of the radioactive tracer and which commonly forms the bulk of the isotopic material. To use relatively large amounts of radioactive material instead of adding carrier may be not only expensive, but also harmful from the intensity of the radiation. For example, to use 0·1 millimole of glucose entirely as ^{14}C—glucose at a specific activity of 20 Ci/mole would need 2 mCi and cost about £500 (Table 2.7) as well as being potentially dangerous. Instead, only a small amount of the ^{14}C—glucose, say 0·1 μCi at a cost of about £0·02, might be used, and made up to the required total mass with unlabelled glucose.

PURITY OF RADIOACTIVE COMPOUNDS

The user of any chemical needs a certain standard of purity which will depend on the use to which the compound is to be put. Problems of ordinary chemical purity apply also to radiochemicals, but in addition it is important to consider the purity of the radionuclide with which the compound is labelled and its correct position in the molecule.

The types of purity as defined in the *Radiochemical Manual*[108] are as follows:

Chemical purity

This is defined as 'the proportion of the material in a specified chemical form regardless of any isotopic substitution', that is, the purity of the chemical without regard to radioactivity.

Radiochemical purity

This is defined as 'the proportion of the total activity that is present in the stated chemical form'. It usually covers both the identity of the radionuclide in the stated chemical material, and its particular location, when specified, in the molecule. For example, the ^{14}C in L−[*ring*−2−^{14}C] = histidine must be located (i) in L−histidine but not in D−histidine nor in any other compound, and (ii) in position 2 of the imidazole ring. Such material is described as *specifically labelled*. Often the radionuclide in labelled molecules is distributed over a number of different positions. It may be distributed uniformly, indicated by the symbol U as in L−[U−^{14}C]histidine, or randomly, when the comparable symbol G is used to indicate that the distribution of the label is general. Tritium compounds are often non-uniformly labelled owing to the loss of ^{3}H from labile positions. The symbol N may also be used to indicate that a label is to a large extent, but not necessarily exclusively, in the designated position.

Radionuclidic purity

This is defined as 'the proportion of the total radioactivity which is in the form of the stated radionuclide'. In the above example ^{14}C should be the only radionuclide present.

Of the three, radiochemical purity is usually of the greatest concern, especially for studies in organic chemistry and biochemistry. Radionuclidic purity is usually adequate for most purposes; for example, the *Radiochemical Catalogue* (1973/4) of the Radiochemical Centre, Amersham, states: 'No radionuclide is normally sold with a purity of less than 99% at the time of despatch; the few exceptions are products which contain two or more radioisotopes of the same element because of the nature of the production process'. High chemical purity of the non-radioactive carrier which is normally present in commercially supplied labelled compounds is not usually important, particularly when the quantity is small relative to the amount of carrier added in the laboratory.

PRACTICAL ASPECTS OF PURITY

It may seem obvious that the measurement of radioactivity in a sample is no evidence of the identity of the labelled compound in it, but this fact may easily be overlooked. At any of the various stages in its use an initially pure radioactive material may become contaminated by other radioactive substances. This may result from degradation during storage, from unexpected chemical change during an experiment or from unintentional introduction of other radioactive material.

The simplest way to judge radiochemical purity, particularly of organic compounds, is by thin layer or paper chromatography. By running a sample of material in several solvent systems a reasonable but by no means certain estimate[8,23] may be obtained of the number of radioactive compounds present and of their identity. Artefacts may, however, be produced by decomposition of the compound during chromatography, or by chemical reaction with chromatographic solvent. Purity may also be assessed by other methods, including isotope dilution analysis[23, 38].

Initial purity

It is advisable to check the purity of commercial material, partly to confirm the statement made by the supplier, and partly as a reference point with which to compare both experimental results and the identity of the stored material at a later date. A reputable source of supply should provide analytical data of purity; for example, the Radiochemical Centre, Amersham, provides an analytical sheet, such as Table 2.8, with each sample of radioactive material delivered.

Purity during use

Radioactive material present at the end of an investigation may not all be in the same form as the radioactive compound originally introduced. Indeed, tracers are commonly used to investigate such changes, but even where no molecular alteration is suspected it is still essential in many cases to check the identity of radioactive material at the end of an experiment.

Purity during storage

Radioactive compounds are particularly liable to decompose during storage, and their purity should therefore always be checked before use.

TABLE 2.8. Technical information provided with material supplied in 1971 by The Radiochemical Centre, Amersham, England (Reproduced with permission.)

L—GLUTAMIC ACID—C14(U)   Batch
Code CFB. 10   analysis
Batch 63   sheet 9387

Protein—C14(U) is obtained from disrupted *Chlorella* cells which have been grown on bicarbonate—C14 as the sole source of carbon. From the acid hydrolysate of this protein, L—glutamic acid—C14(U) is isolated by ion-exchange chromatography, and further purified by paper and ion-exchange chromatography.

Additional evidence for the uniformity of labelling of the amino-acids produced from *Chlorella* has been obtained by ninhydrin decarboxylations see report by HALLOWES, K. H., *et al.*, *Nature, Lond.*, 181, 336, 1958.

BATCH TECHNICAL DATA

Specific activity   :   10·0 mCi/mmol
                                                68·0 μCi/mg

Molecular weight   :   147

Radiochemical purity

    by dilution analysis with L—glutamic acid   :   99%

    by paper chromatography in
      (a) *n*—butanol: water: acetic acid (120:50:30)   :   99% (97%) *
      (b) *n*—butanol: pyridine: water (1:1:1)   : > 98% (97%) *
      (c) ethanol: ammonia: water (80:4:16)   :   −   (97%)*

    by paper electrophoresis
      (a) at pH 2·6   :   98%
      (b) at pH 10·6   : > 98%

Analysed November 1969   * Most recent re-analysis June 1971

Packaging and Storage

L—Glutamic acid—C14(U) is supplied freeze-dried, sealed under nitrogen either in multidose glass vials or in borosilicate glass ampoules.

An ampoule may be opened by lightly scratching the serration with the file provided and carefully snapping off the upper end.

According to available information, the rate of decomposition of this compound should not exceed 1% per year when stored at 2°C, the temperature at which stocks are held in our laboratories.

The purity of all our labelled compounds is determined by our Quality Control Department at carefully chosen intervals.

CAUSES OF DECOMPOSITION DURING STORAGE

Changes in the identity of a radioactive compound may occur during storage for a number of reasons, and may be due to chemical or physical effects unrelated to radioactivity or may be primarily due to the presence of radioactivity.

Decomposition caused by chemical or simple physical factors

Alteration of a radioactive compound may occur as a result of chemical reaction of the molecule with impurities in its neighbourhood or as a result of unsuitable physical conditions of storage. A knowledge of the chemical properties of the compound whose radioactive form is being stored should suggest the likelihood of decomposition and the preventive measures required.

Decomposition resulting from bacterial contamination

Many materials are subject to decomposition as a result of contamination with bacteria. This may be reduced by storing dry at deep-freeze temperatures or under sterile conditions. The introduction of bacteria is particularly likely to occur during repeated withdrawal of samples from a stock solution, and it may therefore be advisable to subdivide the radioactive material for storage.

Decomposition due to the presence of radioactivity (radioactive decomposition)

Radioactivity itself may cause decomposition in three ways[5]. These are as follows:

1. *Internal primary radiation effect*

There is an inevitable change in the identity of a labelled molecule resulting from the decay of a constituent radionuclide to a different and usually non-radioactive product (for example, $^{14}C \rightarrow {}^{14}N$). Furthermore, the disintegration releases energy and affects the charges of the atoms concerned, and these processes have a disruptive effect on the molecule in which the disintegration occurs. The consequences depend on the number of radioactive atoms incorporated into each molecule. If there is only one radioactive atom in each, the resultant products are non-radioactive and are not therefore detected; if there is more than one, some of the products appear as radioactive contaminants. Thus a glucose molecule labelled with a single atom of ^{14}C is converted into a species which is no longer glucose and no longer radioactive. The quantity of products of this type is usually chemically negligible. If, however, two atoms of ^{14}C were present in the glucose molecule, the decay of one would result in a product which, although no longer glucose, would still be radioactive. The rate of appearance of radioactive impurities by this means depends on the rate of decay and on the molar specific activity of the labelled compound as prepared; interference depends on the proportion of radioactive impurities to total radioactive material. This type of decomposition is seldom of great importance except in the case of macromolecules when each contains large numbers of radioactive atoms[108]. Its effect can be avoided by the use of labelled compounds which contain only one radioactive atom per molecule (specifically labelled material).

2. External primary radiation effect

This is the alteration in the molecular structure of a compound as a result of interaction with a nuclear emission originating outside the molecule. If the molecule so altered is radioactive, it will give rise to a radioactive impurity, but the chief importance of this effect lies in the secondary radiation effect which may follow.

3. Secondary radiation effect

This is the chemical change in a molecule interacting with a chemically reactive species, such as a free radical, created by the primary radiation effect. It is thus caused only indirectly by radioactivity. Again, any radioactive molecule affected may become a radioactive contaminant. This effect is responsible for most of the decomposition of stored radioactive materials and may be accelerated by the presence of impurities.

CONTROL OF DECOMPOSITION

Decomposition increases as specific activity increases. Furthermore, radionuclides of low energy of emission decompose more readily than do those of high energy, since the energy of the former is usually expended entirely on molecules within their immediate neighbourhood. Thus, ^{3}H–labelled compounds, with their usually higher specific activity and more localised dissipation of energy, decompose more readily than ^{14}C–labelled compounds. As an example of the rate of decomposition that can occur, most ^{14}C–labelled compounds show a deterioration of less than 3% in a year, whereas the rate of decomposition of ^{3}H–labelled compounds can be much greater. Deterioration of less than 1% in a year ([2–^{3}H]–glycine, 176 Ci/mole) to as much as 80% in less than a month ([G–^{3}H] 9, 10–dimethyl–1, 2–benzanthracene, 14 500 Ci/mole) have been reported for ^{3}H–labelled molecules[8].

The problem of decomposition is complex, and more details are given elsewhere[5-8,32].

Decomposition may be reduced by dilution, by dispersion and by storage under special conditions[8], as described below.

Dilution

Radioactive material may be diluted with carrier, or, if it is required carrier-free, with a suitable diluent (for example, ethanol) which will absorb radiation energy without transferring it or which will combine with reactive radicals. The diluent chosen should be one that may be readily removed so that the radioactive material is restored to its original specific activity. Water is not a satisfactory diluent, since it produces reactive species, such as hydrogen radicals and peroxides. Purity of the diluent is essential in view of the possibility of chemical reaction of the labelled molecules with impurities.

Dispersion on a solid surface

Decomposition may be reduced by spreading the radioactive material over a large surface, such as filter paper, and eluting when required. The surface must be clean and pure, for impurities may react with the labelled compound to produce a labelled impurity. The radioactive material should be dehydrated whenever possible, since small amounts of water may accelerate decomposition. Dispersion is relatively less effective for ^{3}H compounds than it is for those containing ^{14}C.

Storage at low temperature

Storage at -20°C or less usually reduces the rate of chemical reaction and hence of chemical decomposition. However, ^{3}H–labelled compounds may decompose more rapidly if their aqueous solutions are stored between $0°C$ and $-100°C$. A temperature of $-200°C$ has been recommended for storage of ^3H-containing compounds.

Storage under an inert atmosphere

The removal of oxygen from the atmosphere, such as by storing in a vacuum or under nitrogen, reduces the risk of oxidation.

Storage under sterile conditions

Where samples are taken repeatedly from a stock solution of bio-degradable substances an aseptic technique is essential to avoid the introduction of bacteria.

The following general suggestions for storage have been made in the *Radiochemical Catalogue* (1973/4) of the Radiochemical Centre, Amersham:

1. Store at the lowest molar specific activity acceptable.
2. Disperse solids as much as possible and store in a dry atmosphere, preferably under vacuum or inert gas.
3. When possible, store as dilute solution in oxygen-free benzene at room temperature.
4. Keep solutions in the dark.
5. Add bacteriostats where appropriate.
6. Add 1–3% ethanol to aqueous solutions as stabiliser.
7. Store aqueous solutions of ^3H-labelled compounds at $2°C$ or (preferably) at less than $-140°C$.

A simple method of storage which is often adequate for many radio-chemicals is the following. Aliquots of a solution of the radioactive material of about the amount needed for each investigation are applied to regions of chromatography paper located by pencil marks, and the solvent removed. The paper is stored under suitable conditions and the material eluted when required. For a material readily soluble in water, about 90% is eluted in 10 minutes; the exact amount is unimportant, since the activity is usually determined subsequently as described in Chapter 7.

For further discussion of the topics mentioned in this chapter *see* References 36, 42, 61, 68, 92, 104, 108.

CHAPTER THREE

The liquid scintillation counter

Instruments used for the quantitative determination of radioactive materials are of three main types: the gas-filled chamber (which includes the Geiger–Müller counter), the semiconductor detector and the scintillation counter. Each is particularly suitable for certain purposes, but it would be out of place to discuss them in detail here, and the reader is referred to other works[28 a, 102, 104]. Table 3.1 gives a comparison of the characteristics of some of these instruments.

For the measurement of many radioisotopes the scintillation counter has the greatest sensitivity, and its introduction has made possible the accurate measurement of very small amounts of a wide range of radioactive materials. In this type of instrument, energy from emitted radiation is absorbed by a fluorescent material (*scintillator* or *fluor*) and re-emitted as light photons; these are detected by a photomultiplier tube and converted to electrical energy for analysis.

Scintillation counters are of two main types: the *solid* (*crystal* or *plastic*) *scintillation counter* and the *liquid scintillation counter.* In the solid scintillation counter, radiation must penetrate a scintillator which is in solid form (commonly a sodium iodide crystal) before it can cause the emission of light photons; this instrument is therefore designed to measure primarily radiations of high energy and penetration such as γ–rays. In the liquid scintillation counter, on the other hand, the radioactive material is brought into close relationship with a scintillator, usually by dissolving both in a suitable solvent; this type of instrument is therefore particularly suitable for the quantitative measurement of radiations which have limited penetrating power, such as α–particles,

TABLE 3.1. Characteristics of some radiation detectors
(From C. H. Wang and D. L. Willis, *Radiotracer Methodology in Biological Science*, 1965. Reprinted by permission of the authors and Prentice-Hall, Inc., New Jersey, U.S.A.)

Detector type	Energy discrimination	Detection medium	Reproducibility	Detector amplification factor	Relative electronic amplification required	Resolving time (μsec)	Relative background	Relative detection efficiency			Most commonly used sample form	Relative degree of difficulty in preparing sample
								α	β	γ		
Ion chamber with vibrating reed electrometer	Yes	Gas	Excellent	1	Very high	Not applicable	Low	Good	Good	Poor	Gas	High*
Proportional detector	Yes	Gas	Good	10^2–10^4	High	5–50	Low	Good	Good	Poor	Solid	Medium**
Geiger-Mueller detector	No	Gas	Fair	10^7	Low	100–1000	Medium	Good	Good	Poor	Solid	Medium**
Solid (external-sample) scintillation detector	Yes	Solid	Good	10^6	Low	0·3–> 1	High	Fair	Fair	Very good	Solid or Liquid	Medium** or Low***
Liquid (internal-sample) scintillation detector	Yes	Liquid	Fair	10^6	Medium	0·001–< 1	Low	Very good	Very good	Fair	Solution	Low

* Usually involves individual combustion of samples to gaseous form. **Usually involves preparation of planchet-mounted samples.
***Where well-type detector used for liquid samples.

β–particles and soft X–rays (Table 3.1). It also has a very short resolving time and high rates of disintegration can be measured.

The liquid scintillation counter has certain disadvantages, however. Ideally, the radioactive sample should be dissolved in one of a limited group of solvents to obtain greatest accuracy; aqueous samples can only be incorporated in the presence of an intermediary solvent, and this

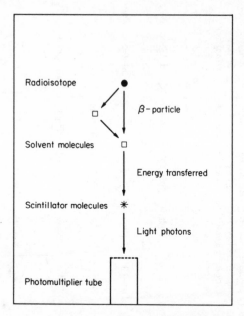

Fig. 3.1. Conversion of energy of β–particle to light. β–particle energy is transferred first to solvent molecules, then to scintillator molecules which re-emit it in the form of light

results in some reduction in sensitivity and accuracy. Gases can be measured only if first trapped in solution. Insoluble materials can be measured in suspension or after being deposited on a solid support, but accuracy is then affected by a number of factors and can vary greatly. Samples are not recoverable after measurement, and so cannot be used for further investigation. Lastly, the efficiency of detection is affected by the presence of contaminating material accompanying the sample, and a correction factor has to be determined for each sample.

Radioisotopes which emit β–particles of low energy or α–particles, but not γ–rays, cannot be measured in a solid scintillation counter, but if γ–rays are also emitted, either instrument may be used. Whenever possible, it is preferable to use a solid scintillation counter, since problems in the preparation of material and the interpretation of data are much less than with a liquid scintillation counter. Some of the most important radioactive tracers used in organic chemistry and bio-chemistry, however, emit β–particles of low energy but no γ–rays (for example ^{3}H, ^{14}C and ^{35}S), and for these only a liquid scintillation counter is suitable.

The measurement of radioactivity by means of a liquid scintillation counter involves two main stages. First, energy from the emitted radiation excites molecules of a suitable solvent (*primary solvent*), which may in turn excite others (Fig. 3.1); from these, energy (in all about 5 per cent of the original β-particle energy[80]) is transferred to molecules of a suitable scintillator (*primary scintillator*) and re-emitted as light (*scintillations*). Second, the emitted light photons are detected by a photomultiplier tube and converted to electrical pulses. If the wavelength of the emitted light does not match the characteristics of the photomultiplier a second scintillator (*secondary scintillator*), which can absorb photons of one wavelength and re-emit at a longer wavelength, may also be needed (*see* p. 70).

Various additional reagents may be added according to need (*see* Chapter 6).

The radioactive material which is to be measured is mixed with a scintillation mixture in a transparent vial, the *counting vial*, which is then lowered into a light-tight *counting chamber* or *well* situated between the two photomultiplier tubes of a liquid scintillation counter. The whole region is surrounded by lead shielding to exclude as much extraneous radiation as possible. Light derived from scintillations in the counting vial enters the photomultiplier tubes either directly or after reflection from the walls of the chamber, whose internal surface is lined with reflecting material.

INSTRUMENT CIRCUITS AND THEIR FUNCTIONS

The functions of a liquid scintillation counter are to detect scintillations derived from radioactive emissions, and to convert the energy of the photons to electrical pulses whose energies reflect those of the original emissions.

Depending on the requirements of the operator, the instrument can be set to select certain pulses, recording some and rejecting others. A pulse which is recorded is called a *count*.

The continuous emission of β–particles from the radioactive material causes the continuous generation of pulses so that the number of counts progressively accumulates. The counts accumulated after any given time are shown either as an illuminated number (*display*) or as a printed number (*printout*). Accumulated counts may be referred to as *the count* or expressed in terms of time as a *count rate*, usually as *counts per minute* (cpm).

Some of the recorded counts are not derived from the radioactive sample. These are called *background counts* (*see* p. 135) and may be due, for example, to the effect of extraneous radiation such as cosmic rays and to pulses generated spontaneously within the electrical circuit. To obtain the count rate of the sample, therefore, the background count rate must be separately determined on a vial containing the same non-radioactive materials as the vial containing the sample and subtracted from the total count rate.

The electrical circuit of the instrument, which is concerned with the detection of scintillations and the processing of the pulses, may be divided into two main parts: a *detection and amplification circuit* and a *pulse analysis circuit*, as shown diagrammatically in Fig. 3.2. These two parts will be described separately, although in practice they often overlap.

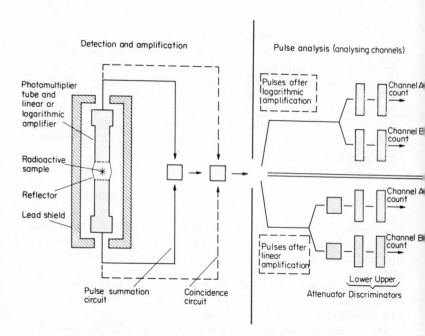

Fig. 3.2. Simplified block diagram of circuit of liquid scintillation counter with linear or logarithmic amplification. An instrument with logarithmic amplification also has a gain control (not shown) in the amplifying circuit

DETECTION AND AMPLIFICATION CIRCUIT

This converts the energy of the scintillation derived from a β–particle into a measurable electrical pulse.

Detection

Two photomultiplier tubes detect scintillations and convert their energy to pulses of current which are then amplified.

A photomultiplier tube is a vacuum tube containing a photocathode and a series of dynodes across which a high voltage is maintained. A photocathode is a light-sensitive surface which emits electrons when light photons impinge upon it. The dynodes are electrodes each of which is kept at a potential which is positive either to the photocathode, in the case of the first dynode, or to the preceding dynode in the case of the remainder. This potential causes electrons emitted by the photocathode in the presence of light to pass to the first dynode and then for each dynode in the series to emit more electrons than it receives. The overall effect is that the photomultiplier generates a pulse of current. This is then amplified and a voltage (*pulse voltage* or *pulse height*) is subsequently developed by passing it through a resistor.

In practice, two photomultiplier tubes instead of one are used to give greater sensitivity and more accurate analysis with a reduction in the background count rate[58]. They are used in conjunction with two additional circuits, as follows:

1. *Coincidence circuit*

A single photomultiplier tube produces a large background count rate consisting of spurious pulses of relatively low voltage generated by the spontaneous emission of electrons from the photocathode. The voltage is similar to that of pulses derived from ^3H, so that this radioisotope cannot easily be measured with only a single photomultiplier tube. The number of spurious pulses can be reduced by cooling of the tube[12, 86], but is still excessive even under refrigerated conditions. However, most of the background count rate from this source is nowadays eliminated by the use of two photomultiplier tubes in conjunction with a coincidence circuit. This rejects any pulses which are not derived from the two tubes at virtually the same time, that is, which are not caused by scintillations. The reduction of background in this way is shown in Fig. 3.3.

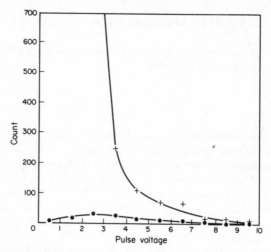

Fig. 3.3. Effect of coincidence circuit on background count rate. Sealed reference background vial counted over different pulse voltage ranges with upper discriminator set 1 V higher than lower discriminator. Values plotted at voltage midway between settings. Crosses represent the count obtained with only one photomultiplier tube in circuit, and circles the count obtained with two photomultiplier tubes in conjunction with a coincidence circuit. Counting time, 10 min. Attenuation zero (gain maximum)

2. Pulse summation circuit

The pulses derived from the two tubes at the same time are combined before being passed as a single pulse to the analysis circuit. This has two advantages:

1. The final pulse voltage is greater as a result of the addition of the individual pulse voltages.
2. Scintillations of the same absolute brightness, regardless of their position in the counting chamber, are converted to pulses of the same voltage. The response of a photomultiplier tube depends on the amount of light impinging upon its face, and this in turn depends on the distance of the light source (that is, the scintillation) from it assuming other factors to be constant. When two tubes are placed opposite each other, a scintillation which is nearer to one will be further from the other, and *vice versa*. Thus for a given brightness of scintillation, the total energy received by the two tubes (and hence the total pulse voltage) should be constant regardless of the distance of the scintillation from either tube.

Amplification

Before they can be observed and analysed, pulses generated by the photomultiplier tubes must be further amplified. However, the maximum energies of β–particles emitted by commonly used radio-isotopes range from about 0·018 MeV (^{3}H) to about 1·7 MeV (^{32}P), and it is technically impracticable to analyse pulses derived from such a large range of energies without modification. Two types of instrument have been designed to overcome this problem.

One type uses *linear amplification*, in which the pulses of current from the photomultiplier tubes are amplified uniformly to the maximum extent feasible for that instrument, and in which the pulse voltage generated is thus *directly* proportional to the energy of the pulse of current. If required, this voltage may then be reduced in the analysis circuit (by attenuation, *see* below) to bring it within a suitable range for measurement, the extent of the reduction depending on the voltage of the amplified pulses.

The other type uses *logarithmic amplification*, in which the pulse voltage is made proportional to the *logarithm* of the energy of the incoming pulse of current, so that the voltages of all pulses are brought within a technically practicable range without further modification. (Occasionally some adjustment in amplification may be required.)

We shall describe mainly the use of instruments with linear amplification; for those with logarithmic amplification the principles are the same except that descriptions of attenuation adjustment may be ignored. Linear amplification is employed by Nuclear–Chicago, Packard and Tracerlab, and logarithmic amplification by Beckman, Intertechnique, LKB and Nuclear Enterprises.

PULSE ANALYSIS CIRCUIT

After leaving the detection and amplification circuit a pulse passes to an *analysing channel*. Each channel has an *attenuation control* (not present with logarithmic amplifiers) which adjusts the pulse voltage, and *discriminator controls* which select certain pulses on the basis of their voltage.

The voltage of each pulse is proportional to the energy of the original β–particle, and therefore pulses show a range of voltages which reflects the original energy spectrum of the β–particles. An analysing channel can be used to pick out a particular part of the pulse voltage spectrum (*pulse analysis*) and hence to examine a chosen part of the β–particle spectrum.

There may be several analysing channels in one instrument, all able to process a pulse independently and simultaneously.

Attenuation

With instruments which use linear amplification, it may be necessary to reduce the pulse voltage to bring it within a suitable range for analysis: this adjustment is made by means of an *attenuator*. If attenuation is

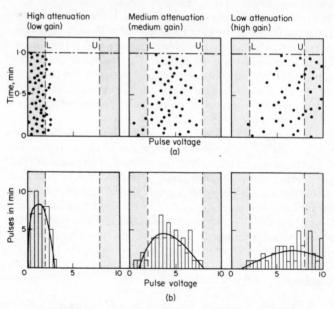

Fig. 3.4. Effect of attenuation on pulse voltage in instrument with linear amplification, and use of discriminators. The diagrams are hypothetical; if the voltage range is arbitrarily made to be 0–10 V, the pulse voltages generated correspond roughly to those which might be derived from unquenched ^{14}C.
   Attenuation
   (a) The dots represent the voltages of the pulses generated over one minute. As the attenuation is increased (gain decreased) (diagrams from right to left) the voltages of the pulses are reduced.
   (b) The pulses in each of (a) are shown in the form of histograms, the vertical bars representing the pulses appearing within equal small voltage ranges during the one minute. The solid curved lines represent continuous pulse voltage spectra drawn through the histogram values.
   Discriminators
   The vertical interrupted lines represent the limits set by discriminators (L., lower discriminator; U., upper discriminator); only pulses in the clear area between the discriminator values (the window) are counted

zero, the pulse voltage is unaltered; as attenuation is increased the pulse
voltage is reduced. The attenuator is often called a *gain control*; high
gain is equivalent to low attenuation, and *vice versa*.

Instruments with logarithmic amplification do not have a control of
this sort in the analysis circuit, but there is usually a gain control in the
amplification circuit which may be used for standardising the instrument
or for altering the overall amplification.

The effect of attenuation after linear amplification is demonstrated
diagrammatically in Fig. 3.4(a), in which each diagram shows the pulses
occurring over a period of one minute. Each point represents a pulse
with the time at which it occurred plotted against its voltage. (The
shaded areas should be ignored at this stage.) The right-hand diagram

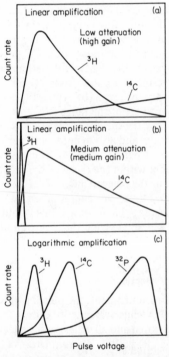

*Fig. 3.5. Spectra of voltages of pulses generated in instruments with linear and
with logarithmic amplification. Diagrams hypothetical.*
   *(a) Linear amplification with attenuation (gain) adjusted for ³H measurement.
No ¹⁴C pulses other than those of lower voltage appear within the pulse voltage
range shown.*
   *(b) Linear amplification with attenuation (gain) adjusted for ¹⁴C measurement.
³H pulses are crowded into the lower end of the voltage range.*
   (c) Logarithmic amplification with low gain in amplification circuit

shows how the voltage of a number of pulses might be distributed with
little or no attenuation, that is, virtually as they arrive from the
amplifier. The pulse voltages are distributed in a way which reflects the
distribution of the corresponding β–particle energies. With increasing
attenuation (decreasing gain) the pulse voltages are decreased (middle
and left-hand diagrams). In Fig. 3.4(b) the number of pulses present at
each voltage level is plotted in the form of a histogram. In each case
a smooth curve has been drawn through the histogram values to give
the *pulse voltage spectrum.*

The shape of a spectrum of this sort within a particular pulse
voltage range depends not only on the attenuation but also on the
β–particle energies of the radioisotope measured. This is shown in
Fig. 3.5 (a and b), where the spectra of ^{3}H (low β–particle energies) and
^{14}C (higher β–particle energies) are superimposed, and shown as they
might appear within the operating range of an instrument with linear
amplification. With low attenuation (a) the spectrum of ^{3}H is suitably
placed within the range, whereas most of the ^{14}C pulses are at higher
voltages (off the diagram to the right). If the attenuation is increased
(gain decreased) the voltages of pulses from ^{14}C are reduced so that
they lie optimally within the range (b), whilst those of ^{3}H are now
crowded into the lowest part of the range.

Fig. 3.5(c) shows how the spectra of ^{3}H and ^{14}C might appear with
logarithmic amplification, and also includes another radioisotope, ^{32}P,
which has much higher β–particle energies. The spectra of the three
isotopes fit comfortably within the range of the instrument without
further adjustment. For the measurement of ^{3}H alone (or any other
radioisotope with emissions of low energy) the gain of the detection
circuit might with advantage be increased; this would extend all the
spectra towards the right and expand the spectrum of ^{3}H.

Discrimination (pulse selection)

After attenuation (gain) adjustment, the pulses within a desired voltage
range may be selected by electrical circuits known as *discriminators*, so
that they alone are counted and all other pulses rejected.

Each analysing channel has two discriminators: a *lower discriminator*
(also called *lower level*, *lower threshold* or *threshold*) which determines
the lower voltage limit of those pulses which are counted, and an *upper
discriminator* (also called *upper level*, *upper threshold* or *window*)
which determines the upper voltage limit. (The meaning of the word
'window' here must not be confused with the more generally accepted
one about to be described.) The counting of pulses between limits set
by discriminators is called *differential counting*, and the voltage range

between the two discriminators is known as a *window*. It may often be preferable, however, to count all pulses above a lower limit set by one of the discriminators, so that the upper limit of the window is effectively infinite volts; this is called *integral counting*. The lower discriminator should not be used below the lowest practicable setting, that is, the lowest setting recommended by the manufacturer, otherwise spurious pulses may be counted; if this setting is used in conjunction with the highest setting of the upper discriminator, the discriminators are set at their widest practicable limits.

A window of reduced voltage range has been superimposed on the other symbols in Fig. 3.4, in which the lower and upper discriminators are represented by two vertical interrupted lines at 2V and 8V respectively. Only pulses within the window, represented by the central clear area, are counted.

The more usual reasons for using discriminators are as follows:

1. To reduce background. The discriminators may be used to exclude background pulses whose voltages are different from those of the pulses being examined.
2. To select particular parts of the pulse voltage spectrum, as for example, (i) in the measurement of the spectral shift associated with energy losses in the scintillation process (*see* 'Quenching', Chapters 4 and 7), and (ii) in the analysis of a mixture of radio-isotopes (Chapter 8).

Rough pulse voltage spectra can be constructed by the use of the discriminators, as shown in Fig. 3.6, which shows spectra of ^{3}H at three attenuator (gain control) settings. With each attenuator setting fixed, the discriminator settings were moved, 1 V at a time, over the complete voltage range, counting each time with the upper discriminator 1 V above the lower discriminator. Histograms have been constructed from the counts obtained and curves drawn through them to represent the spectra. They are comparable with the hypothetical histograms and spectra shown in Fig. 3.4(b).

Balance point

If attenuation (gain) is progressively changed, the pulse voltage spectrum will move across a window which has fixed limits, as can be seen in Figs. 3.4 and 3.6. It should be clear from the figures that there is an attenuator setting at which a maximum number of pulses lie within the window, that is, at which the count rate for those discriminator settings

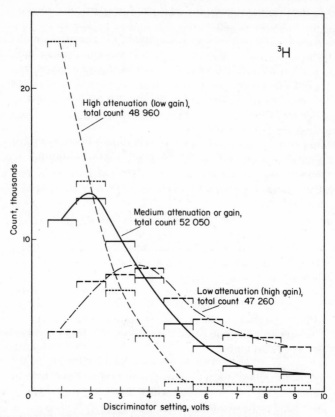

Fig. 3.6. Effect of attenuation (gain) on pulse voltage spectrum of ^{3}H. At each of three attenuation settings the discriminator settings were moved, 1 V at a time, over the voltage range 0·5–9·5 V, and the upper discriminator was set 1 V above the lower discriminator. The counts have been plotted in histogram form and approximate continuous spectra drawn by eye through the values obtained. Counting time, 1 min. 'Total count' is the count obtained from the whole range, 0·5–9·5 V.

Source: ^{3}H–toluene Amersham/Searle sealed reference standard, 85 500 dpm

is greatest. This setting is known as the *balance point*. In both figures, the balance point corresponds to a setting giving medium attenuation.

The balance point concept does not apply to instruments with logarithmic amplification because there is no control for attenuation or gain in the analysing channels.

With differential counting, that is, when counting pulses whose voltages lie within limits set by discriminators, the use of a balance point setting is advisable for the following reasons:

1. It provides the maximum count rate for a particular sample within a chosen window.
2. It reduces error caused by instability of the electrical circuits. A change in amplification in the detection circuit causes a shift of all pulse voltages and therefore causes a change in count rate. This change is least at the balance point, since the rate of change of any property is least in the vicinity of a maximum or a minimum.

The balance point may change if the conditions of counting are varied in any way, particularly by alterations to the following:

1. Discriminator settings

The attenuator (gain control) at balance point places a pulse voltage spectrum optimally within a particular window. If the window is changed, it should not be assumed that the balance point of the new window will be the same as that of the old.

2. The radioisotope being measured

Different radioisotopes emit β–particles of different mean energies and hence give different pulse voltage spectra. It follows that each must require a different setting of the attenuator (gain control) to fit the pulses optimally within a given window. This property may help to identify a radioisotope.

3. The scintillation mixture

The constituents of a scintillation mixture can affect the pulse voltage spectrum (*see* 'Quenching', Chapter 4) and thus the position of the balance point.

4. Performance of liquid scintillation counter

The balance point setting depends on all factors associated with the measurement of scintillations, that is, on mains voltage, overall amplification, discriminators, attenuators and other components. It may therefore vary in different instruments, in different channels of the same instrument, or even in the same channel of an instrument at different times. It should therefore be checked frequently, and *always* after servicing or repair.

CHOOSING THE BEST SETTINGS FOR COUNTING

When measuring a radioactive sample, it is obviously desirable to obtain the highest possible count rate with the least interference from background. The background count rate is affected by the instrument settings, and the larger the voltage range over which the sample is counted, the greater will be the contribution from the background. It follows that counting within a limited voltage range, that is, within a window (differential counting) will give a lower background count rate than counting over an unlimited voltage range (integral counting). There are thus two main ways of counting a sample, depending upon the need to reduce background.

1. Maximum integral count

If it is unnecessary to count within a window, a maximum integral count may be obtained. (This is sometimes, but inaccurately, called 'integral count'.) For this, all the pulses reaching the analysis circuit are counted, and consequently the background count rate is higher than with a maximum differential count (*see* below). This method is thus not suitable for counting samples whose count rate approaches that of the background.

Instrument settings are as follows:

1. Set lower discriminator at its lowest practicable setting.
2. Make the upper discriminator inoperative, that is, count all pulses whose voltages lie between that of the lower discriminator and infinity (integral counting).
3. Set attenuator at zero (gain maximum).

For a sample whose count rate is high enough for the resulting raised background to be acceptable, maximum integral counting has several advantages over counting at the balance point. These advantages are:

1. The count rate is the maximum obtainable for the sample being counted.
2. The instrument settings are the same for all samples, all instruments and all times, and do not need to be determined experimentally.
3. The stability of counting is greater, since a potential source of instability, that is, the upper discriminator, is eliminated from the circuit.
4. The same settings on all analysing channels should provide an identical count rate for a given sample in a particular instrument.

2. Balance point

If the radioactive material must be counted within a window, the balance point gives the highest count rate for the discriminator settings used.

The orthodox method of determining the balance point is to vary the attenuator (gain control) until the count rate within the chosen window reaches a maximum. Owing to the random nature of radioactive decay, this may take some time, since it requires a high total count for reliability.

A more rapid method[72] eliminates the random element by comparing the count rate within the window of the analysing channel whose balance point is required (say Channel A), to the count rate in a second channel (say Channel B) which is set to count *all* pulses (maximum integral count). The balance point is then that setting of the attenuator (gain control) for which the ratio of the count in Channel A to the count in Channel B is a maximum. The procedure is as follows (2, 3 and 4 are primarily included to show the principle; 5 is the method of choice).

1. Initial instrument settings. Channel A (whose balance point is to be determined): set discriminators as required; attenuator (gain control) setting initially arbitrary. Channel B: set for maximum integral count.
2. Count for a time which gives, say, about 4000 counts in Channel B. Note count in both channels.
3. Calculate ratio of count in Channel A to count in Channel B.
4. Repeat (2) and (3) with progressive alteration of attenuator (gain control) setting of Channel A (Fig. 3.7 (c) and (b)). The balance point is the attenuator (gain control) setting which gives the highest ratio.
5. For a more rapid determination, set instrument to stop when each count in Channel B reaches a chosen value (say, about 4000 counts). The denominator of the ratio is now constant, and the balance point is thus the attenuator (gain control) setting of Channel A at which the count in that channel is a maximum (Fig. 3.7 (a)).

Counting at the balance point with the discriminators at their widest practicable limits gives the *maximum differential count*. This is the maximum count rate obtainable when both discriminators are in use.

For instruments with logarithmic amplification, where the balance point concept does not apply, the background count rate can be minimised in the following way. For a chosen gain in the amplifying circuit adjust the discriminators so that the pulse voltage spectrum just lies between their limits.

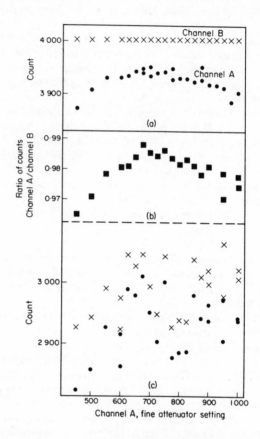

Fig. 3.7. *Determination of the balance point. Two analysing channels (A and B) are prepared as follows. Channel A (whose balance point is to be determined): discriminators set at widest practicable limits (or other settings chosen). Channel B set for maximum integral count. Counts in Channel A represented by* circles; *counts in Channel B represented by* crosses.

(a) Rapid determination of balance point. Instrument set to stop at 4000 counts in Channel B. Balance point indicated by maximum count in Channel A.

(b) Ratio of counts in Channel A to those in Channel B (from (c)). Balance point indicated by maximum ratio.

(c) Counts in both channels measured for 1 min at various attenuation (gain) settings.

Source: ^{14}C–*toluene Amersham/Searle sealed reference standard, 31 000 dpm*

THE LIQUID SCINTILLATION COUNTER

45

Programmed channel settings

Some instruments are so designed that it is unnecessary to set the
discriminators individually. Instead, a set of plug-in components (or the
equivalent) is provided, each of which automatically sets the discriminators (and attenuators, where applicable) to match the pulse voltage
spectrum of the radioisotope to be measured; the settings may also be
modified depending on the degree of quenching expected.

INSTRUMENT CONTROLS

The multitude of controls on a liquid scintillation counter may appear
formidable at first, but once the purpose of the various sections of the
electrical circuit have been understood, the use of the controls should
be clear.

Controls may be divided into three main groups: those for the
positioning of the counting vial and external standard, those for the

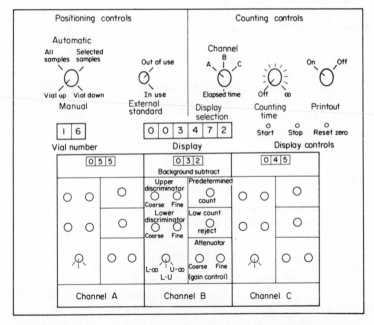

*Fig. 3.8. Hypothetical instrument panel of automatic liquid scintillation counter
with linear amplification. An instrument with logarithmic amplification would
have a single gain control, instead of an attenuator (gain control) for each channel*

counting itself, and those for pulse analysis. A simplified control panel for an instrument with linear amplification is shown in Fig. 3.8; an instrument with logarithmic amplification would differ only in having one separate gain control in the amplification circuit instead of an attenuator or gain control for each channel.

Positioning controls

The *sample-changer* provides a choice, in most instruments, between manual and automatic counting, that is, a single chosen vial may be moved into the counting chamber or each vial of a group may be counted in turn automatically.

An *external standard control* brings a γ-emitting radioactive source (usually ^{226}Ra, ^{133}Ba or ^{137}Cs) close to the counting chamber when required (*see* Chapter 7).

Counting controls

These are of three main types:

1. A *timing control*, which selects the time for which a sample is to be counted
2. A *display control* which determines whether the accumulated count in any one of the analysing channels or the elapsed time is to be shown.
3. A control for operating a *printer* or *lister*.

Printout

The form in which data are printed may vary from one instrument to another, but the following is common:

16	(vial number)
2·00	(counting time, minutes)
3472	(count in Channel A)
4715	(count in Channel B)
21 256	(count in Channel C)

Other information may be added, for which the instruction manual of the instrument should be consulted.

Analysing channels

There may be several analysing channels, so that a radioactive sample may be counted with different groups of settings for pulse analysis at the same time. Identical settings on different channels do not necessarily produce exactly the same count rate from a given sample, owing to differences in the properties of similar components.

The controls for all channels on a particular instrument usually have the same layout. Each has an *upper* and a *lower discriminator control* and, with a linear amplifier, an *attenuator* (or *gain control*) with fine and coarse adjustments. The coarse attenuator control often has a limited number of positions for design reasons; the fine control is usually continuously variable and covers a greater range than the limits of each step in the coarse control in order to provide overlap. The $L-U$ *control* limits the count to pulses whose voltages lie between the discriminator settings (differential counting). The $L-\infty$ and $U-\infty$ *controls* provide for the counting of all pulses whose voltages are above the value set by the lower or upper discriminator respectively (integral counting). (Lettering may vary on different instruments; for instance, L–U may be replaced by A–B, and $L-\infty$ by $A-\infty$, etc.)

A control to stop the instrument after a *predetermined count* (*preset count, precount stop*) is usually provided for each channel. It may avoid unnecessarily long counting times, and is particularly useful for a series of samples which have great differences in activity; it is also useful in the rapid determination of balance point and in finding the settings for channels ratios (*see* Chapter 7).

A *low count reject control* stops the counting of a sample whose count in, say, the first minute is below a chosen value, and may be used to avoid the prolonged counting of samples whose low count rate is of little significance.

Synchronous timing

A liquid scintillation counter is only capable of subdividing time into limited fractions of a minute. If the time at which a predetermined count is reached does not coincide with one of these fractions, the next whole time interval will be printed. If the counting time is short the calculated count rate may then be in error, and if the count rate is high, this error may be large. For example, if the shortest time that can be recorded is 0·01 min, if the true count rate is 50000 cpm, and if the instrument is set to stop at 100 counts, the count rate would be given as 100 counts in 0·01 min, that is, 10000 cpm. A *synchronous timing*

control enables counting to continue until the end of the next complete
time interval for that instrument, so that, in the above example, counting would stop at 500 counts instead of the limiting 100 set on the
preset count control. The count rate calculated would then be 500
counts in 0·01 min, giving the correct value of 50000 cpm.

Resolving time

The time between one incoming pulse and the next may be shorter than
the time required to process a pulse. If special precautions were not
taken, two pulses would then be counted as one (*coincidence loss*) and
their voltages would be additive, that is, a single pulse would be recorded
whose voltage would be higher than that of either pulse alone. This is
avoided by automatically preventing the input of further pulses during
the time required to process the first one. This processing time is called
the *resolving time*, but it may also be called the *dead time*, which
implies the time during which incoming pulses are blocked; the time
during which incoming pulses are acceptable is then called the *live time*.
The live time is the actual counting time indicated on the instrument,
and the total operating time is often slightly greater than this, since it
consists of live time plus dead time. For example, we have found that
a count of 470 000 in a counting time of 1 min actually took 1 min 7
sec. Although this discrepancy is of no practical importance, we
mention it to avoid consternation if it is discovered that instrument
time does not agree with clock time.

Caution

The life of a photomultiplier tube may be shortened by putting a high
voltage across it before the cathode has warmed up. If there is no
automatic delay, the high voltage switch (with positions usually *OFF*,
ON and *H.V.*) should not be switched from *ON* to *H.V.* until several
minutes after switching on the instrument. This precaution is not
mentioned in some instruction manuals.

It is preferable to leave a liquid scintillation counter switched on at
all times, since this maintains stability and reliability.

For further discussion of the topics mentioned in this chapter, *see*
References 58, 80, 83, 94, 102, 104.

CHAPTER FOUR

Counting efficiency and quenching

The number of counts recorded by a liquid scintillation counter is always less than the number of β–particles emitted in the sample. This discrepancy is usually expressed in terms of *counting efficiency*, which is the percentage (or fraction) of decaying nuclei actually counted.

Efficiencies may vary enormously, depending on the radioisotope and on the counting conditions, but for ^{14}C are usually in the region of 70–95%, and for ^3H in the region of 10–60%.

Counting efficiency depends on a number of factors which are described in detail below. Some factors are associated with the instrument itself, and of these a number are often ignored because they are assumed to be constant, although for precise measurements, as will be seen, they must be taken into account. Other factors are associated with the sample to be counted, and of these the most important is *quenching*, which is the reduction of pulse voltage by material present in the sample or scintillation mixture. In the following list of factors which affect efficiency we have looked upon it in its broadest sense by including any factor which may alter the relationship between disintegration rate and count rate.

FACTORS RELATED TO INSTRUMENT DESIGN

The performance of a particular liquid scintillation counter varies from time to time and should be checked frequently by means of a sealed radioactive standard. Indeed, the count rate derived from such a

standard should be determined routinely with each series of measurements. A check of this sort is nevertheless of limited value, since certain
types of sample may show a variation not shown by these standards
(*see* 'Instrument Stability' below).

Efficiencies claimed by manufacturers are usually reliable, but
naturally refer to measurements made under the best possible conditions
and are often much higher than those found in experimental work.

Instrument sensitivity

The sensitivity of a photomultiplier tube to the photons emitted by
scintillator molecules is of prime importance in determining the proportion of β-particles which give rise to pulses. However, if a photomultiplier tube does not respond to a scintillation, this may not be
because the tube lacks the sensitivity to detect a scintillation of that
brightness, but because the peak wavelength of the emitted light is not
that to which the tube is most sensitive. In this case, the defect may be
overcome by use of a secondary scintillator.

An increase in the voltage across the photomultiplier tube or an
increase in amplification increases sensitivity only up to a certain limit,
above which the increase in electrical 'noise' leads to intolerable
interference. The photomultiplier voltage is nowadays usually set by
the manufacturer and is not under the control of the operator.

Linearity of response

In colorimetric work, it is important for the measured absorbance of a
sample to be proportional to the concentration of the substance being
measured. In liquid scintillation counting, the same principle applies,
that is, the count rate of a sample of radioisotope should increase in
proportion to its rate of disintegration and hence to its concentration.
If very high count rates are to be measured it is therefore advisable to
check linearity up to the maximum count rate expected. An example
of such a check is shown in Fig. 4.1; within the limits of error, the
relationship between count rate and concentration of radioisotope was
found to be linear up to at least 1 500 000 cpm.

In a check of this sort it is essential that all vials contain identical
volumes of fluid and that the radioactive material does not itself cause
quenching in the amounts used, since both these factors affect count
rate. These sources of error can be avoided by making all samples up to
a constant volume with toluene, and by using a non-quenching standard
such as ^{14}C-toluene.

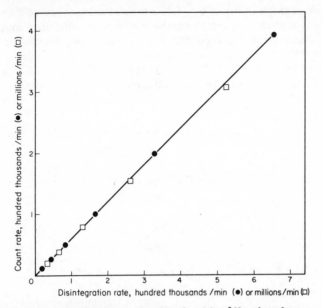

Fig. 4.1. Relationship between count rate and activity. ³H–n–hexadecane was counted in 10 ml toluene containing 8 g PPO/litre and 0·2 g POPOP/litre. Two sets of values are superimposed, one with an upper limit of about 400000 cpm (circles), the other with an upper limit of about 3000000 cpm (squares). Volume of hexadecane was about 1 μlitre/3000 dpm, but unlabelled n–hexadecane showed no quenching at the largest volume used

A rapid method of testing the linearity of response up to a chosen limit is as follows. An amount of radioactive standard which will give a count rate equal to the upper limit required is dissolved in scintillation mixture and then counted. Half the contents of the vial is then transferred to a clean vial and made up to the same volume as the original with the same scintillation mixture. A count rate in this vial of exactly half that of the original indicates that the response is linear.

Instrument stability[111]

The efficiency of counting can be affected by instability of the photomultipliers or of the electrical circuits; this can be caused, for example, by changes in the mains voltage or in the temperature of components.

Instability appears to be more noticeable with low counting efficiency, as when the samples being counted are severely quenched or contain radioisotopes which emit β–particles of low energy. An example of this is shown in Fig. 4.2, in which the count rate of $^{14}C-n-$hexadecane is seen to be more variable the lower the efficiency with which it was counted. It follows that the standards used for checking stability should be counted with the same efficiency as the samples which are being measured. ^{14}C and ^{3}H show a similar degree of variability when counted with the same efficiency.

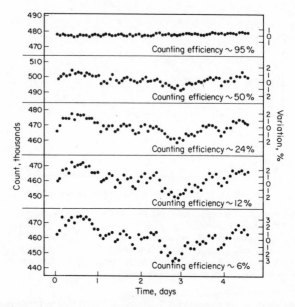

Fig. 4.2. Stability of counting at different counting efficiencies. $^{14}C-n-$hexadecane was dissolved in various quantities of toluene and chloroform (as quenching agent) in a series of glass counting vials, together with sufficient PPO to give a final concentration of 5 g/litre. The total volume of mixture in all vials was the same. The quantities of radioisotope and chloroform were adjusted to give different counting efficiencies but similar count rates. Samples were counted for 20 min each in turn. Instrument set for maximum integral count

When counting efficiency is low, it is advisable to count standards at the beginning, middle and end of a series of samples of unknown activity. If the counting time of such a series must be long, the whole series should be counted repeatedly for short periods, and if no instability is found from one set of counts to another, the values for each sample are added together.

Instability of the instrument is a particularly serious problem in the counting of two radioisotopes in the same sample at low efficiency (Chapters 8 and 9).

Temperature[86]

With an instrument whose counting chamber is refrigerated, time may have to be allowed for a sample vial to cool before counting, since temperature can affect efficiency. This may be unimportant for samples measured at high efficiency, but as shown in Fig. 4.3, the effect may be more marked with samples counted at low efficiency. Here a delay of about 30 min would have to be allowed with some of the samples.

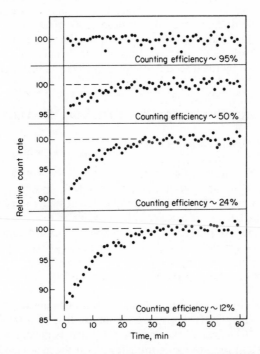

Fig. 4.3. Effect of cooling on count rate of samples counted with different efficiencies. Samples prepared as described in caption to Fig. 4.2, using $^{14}C-n-$ hexadecane as source. Sample initially at 22° C placed in counting well maintained at a constant temperature of about 5° C and counted immediately for 1 min and then repetitively. Instrument set for maximum integral count

Discriminator and attenuator (gain control) settings

The proportion of pulses which are counted depends on the width of the counting window and on the position of the pulse voltage spectrum in relation to the window.

Altering the settings of a discriminator may have little or no effect on the count rate or may affect it greatly, depending on the number of pulses near the voltage level of that discriminator. For example, in the left diagram of Fig. 3.4 (a), a small change in the setting of the lower discriminator would markedly affect the count rate (and hence the efficiency); on the other hand, a much larger change in the setting of the upper discriminator would have little or no effect.

Changes in the attenuation (or gain) may affect the count rate by altering the voltage of all pulses, thus shifting them into or out of the range selected by the discriminator settings, as also illustrated in Fig. 3.4.

FACTORS ASSOCIATED WITH VIAL CONTENTS

Volume of fluid in counting vial

The count rate may vary with the volume of vial contents[88]. The maximum count rate is given by the volume which positions the fluid optimally in front of the photomultiplier tubes. With most instruments this is about 10–15 ml. Smaller or larger volumes than the optimum will therefore give a reduced count rate. Because of differences in transparency and reflection, the effect when using glass vials is different from that with plastic vials; for example, we have found that the effect of a change of volume in the 10–20 ml range is less with polyethylene vials.

Examples of the effect of volume on count rate are shown in Fig. 4.4. Decreasing the efficiency of counting (such as by measuring ^{3}H instead of ^{14}C) increased the sensitivity to changes in volume, because of the greater proportion of scintillations near the margin of detectability. The count rate from the external standard (in this case beneath the vial) was even more dependent on volume, because the greater the volume of fluid the greater the number of molecules which can be excited by the external standard.

β–particle energy

Radioisotopes whose β–particles have energies comparable with those of ^{14}C (0·159 max. MeV) can usually be counted with an efficiency of

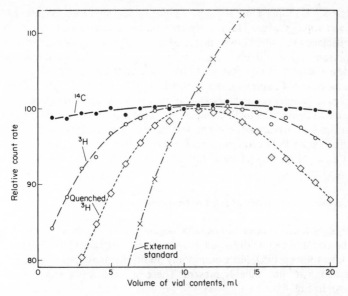

Fig. 4.4. *Effect of volume of vial contents on count rate.* ^{14}C– *or* ^{3}H–*n–* *hexadecane was measured into an empty glass vial (radioisotope omitted when counting external standard) and cooled to instrument temperature (about 5° C) before counting. Scintillation mixture at 5° C was added 1 ml at a time and the sample counted for 1 min after each addition. Relative count rate of 100 represents the following: for* ^{14}C, *85 860 cpm, counting efficiency 96%; for* ^{3}H, *88 120 cpm, counting efficiency 55%; for quenched* ^{3}H, *37 340 cpm, counting efficiency 8%; for external standard, 343 030 cpm. Scintillation mixture: toluene containing 8 g PPO/litre; for quenched* ^{3}H, *7 volumes of this mixture to 1 volume of chloroform. Instrument set for maximum integral count*

over 90% under optimum conditions. However, when the energy of the β–particles is much less, the efficiency of counting is correspondingly lower since a greater proportion of β–particles are now below the level of detection. This is the case with ^{3}H (0·018 max. MeV), which cannot usually be counted with more than about 60% efficiency. (Efficiencies when measuring α–particles, however, may be as high as 100%.)

Quenching

Anything which interferes with the conversion of β–particle energy to scintillations or with light transmission reduces the amount of light which reaches the photomultiplier. As a result, fewer scintillations are detected, and the pulses generated are of lower voltage. Consequently efficiency is reduced and the whole pulse voltage spectrum is shifted to a lower voltage range. This interference (and the consequent reduction in efficiency) is called *quenching*, and is probably the greatest problem in present-day scintillation counting, particularly of biological materials. Unless the quenching is known to be constant or the efficiency can be estimated (*see* Chapter 7), the comparison of count rates derived from different samples is meaningless. Various causes of quenching are discussed below; the most common are substances introduced into the scintillation mixture either to make the sample miscible with the primary solvent or as an unavoidable constituent of the sample.

Causes of quenching are as follows:

1. Certain constituents of scintillation mixtures

The highest efficiencies can only be obtained if the radioactive material to be counted can be dissolved in a simple solvent/scintillator mixture such as toluene/PPO. More commonly additional materials which cause quenching are unavoidably present. These may act in two ways, as shown in Fig. 4.5.

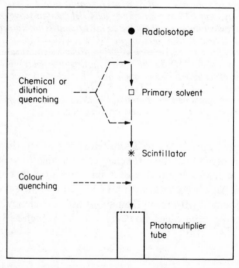

Fig. 4.5. Quenching. The diagram shows the sites of action of quenching agents

(a) By interference with the transmission of energy from β–particle to scintillator.

Some of the energy may be absorbed by an interfering substance (*chemical quenching*) and dissipated as heat; types of aliphatic compounds which act in this way are ketones, amines and halogenated compounds[11, 62]. (Chloroform and carbon tetrachloride are often used as *quenching agents* for deliberate quenching.) Alternatively some of the energy may be prevented from reaching excitable molecules by the diluting effect (*dilution quenching*) of compounds such as aliphatic ethers, esters and alcohols[11, 62]. The overall effect is the same in both cases, that is, a reduction in the energy reaching the scintillator, but the result of chemical quenching is the more marked. It should be noted

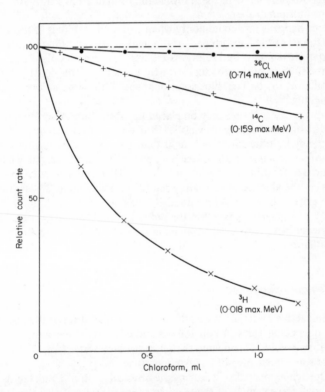

Fig. 4.6. Effect of quenching on the counting of radioisotopes emitting β–particles of different energies. All samples initially prepared in 20 μl of 0·14 M NaCl, 1 ml of cellosolve and 9 ml of toluene containing 4 g PPO/litre. Chloroform was then added in the amounts shown and the volume made up to 12 ml with toluene. Relative count rate of 100 represents the following: for ^{36}Cl, 197000 cpm, counting efficiency not known, but probably greater than 90%; for ^{14}C, 103000 cpm, counting efficiency 94%; for ^{3}H, 36220 cpm, counting efficiency 44%. ^{36}Cl present as Na^{36}Cl, ^{14}C as ^{14}C–n–hexadecane and ^{3}H as ^{3}H–n–hexadecane. Counting time 1 min. Instrument set for maximum integral count

that oxygen[94], water[56], and acids and alkalis are strong quenching agents.

The effect of identical amounts of a given quenching agent on efficiency is greater with β—particles of low energy (e.g. from ^{3}H) than with those of higher energy (e.g. from ^{14}C). This is shown in Fig. 4.6, which shows that amounts of chloroform which have a marked effect on the count rate of ^{3}H have relatively little effect on that of ^{14}C and almost no effect on the count rate of ^{36}Cl.

(b) By interference with the transmission of light from scintillator to photomultiplier tube.

This happens most commonly when coloured materials in the solution to be counted absorb part of the light emitted by the scintillator (*colour quenching*). Red solutions are the most potent colour quenching agents, since they absorb most strongly the wavelengths emitted by the scintillator. In contrast, blue solutions cause little or no colour quenching[100].

In this category also may be placed the absorption of emitted light by dirt or other marks on vials. Plastic vials also reduce counting efficiency by transmitting less light than glass ones; the magnitude of the effect depends on the initial energy of the β—particles, and we have found that for minimally quenched ^{14}C the efficiency with polyethylene vials (in the absence of secondary scintillator) was about 2% less than with glass vials, whereas for minimally quenched ^{3}H it was about 8% less. When quenching is severe the reduction can be even greater. The efficiency with plastic vials may be improved by using a secondary scintillator.

2. Self-quenching

A radioactive material may itself cause quenching if it has the properties of a quenching agent. An obvious example is radioactive chloroform, but it should be remembered that even tritiated water (^{3}H$_2$O) will cause some quenching of its own β—particle energy.

Quenching may also be due to the physical form of a radioactive material. For example, if the material is not completely dissolved, some of the energy of the emitted β—particles will be absorbed by the solid material before it can excite the surrounding molecules of primary solvent. This is an example of the phenomenon known as *self-absorption*, and is particularly evident with ^{3}H—labelled compounds, whose β—particles have limited penetration. It may thus be impracticable to count such compounds in the form of a suspension, since the reduction in efficiency is too great and too variable for accurate measurement.

3. Surface adsorption

If molecules of a radioactive material are adsorbed on to the surface of a counting vial or on to other solid material in the vial, the number of scintillations produced from those molecules may be reduced by as much as half. A radioactive molecule which is dissolved in the scintillation mixture is completely surrounded by fluid, so that the solid angle subtended at the molecule is 4π. However, when it is adsorbed on to a flat surface it is then in contact with fluid on one side only, subtending a solid angle of 2π, so that the energy of only about half the β—particles is transmitted directly to the scintillation mixture. The first condition is often referred to as 4π (or *4pi*) *counting*, while the second is known as 2π (or *2pi*) *counting*.

Homogeneity of vial contents

The vial contents, that is, the liquid scintillation mixture and the sample, may consist of two normally immiscible solvents, such as toluene and water, combined by means of a third, such as cellosolve, which is miscible with both. The mixture is homogeneous provided that the proportions are suitable; if they are not, and especially in a cooled counting chamber, the mixture separates into two phases. The radioactive material then becomes unequally distributed between these two phases, and the count rate of a water-soluble substance falls because of the separation of the radioactive molecules from the primary solvent and scintillator molecules. Separation of this sort can be readily seen in glass vials, but may be overlooked in plastic ones.

Precipitation during counting

In some cases, radioactive materials which appear to be in solution in the scintillation mixture slowly precipitate. Although the precipitation may not be visible to the naked eye, it appears as a slow fall in count rate, and can usually be distinguished from other causes of a fall in count rate by a return towards the original value if the vial is shaken. The fall in count rate may be due partly to sedimentation of particles, and partly to self-absorption by the clumping of precipitating material[64]. An example in which ^{3}H—inulin behaved in this way is shown in Fig. 4.7.

Variations peculiar to plastic counting vials

The materials of which plastic counting vials are made may be permeable to some of the solvents and standards used in liquid scintillation count-

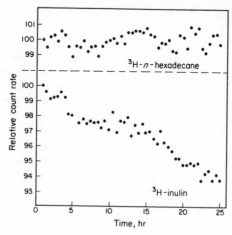

Fig. 4.7. Progressive fall in count rate due to clumping and precipitation during counting. The count rate of ^3H–inulin and ^3H–n–hexadecane (as control) measured every 30 min is shown in relation to the initial count rate (relative count rate of 100). On shaking the vials subsequently, the count rate for ^3H–inulin rose, whereas that of ^3H–n–hexadecane did not. Both materials were counted in identical mixtures in polyethylene vials: 10 ml toluene/cellosolve (1:1 v/v) containing 4 g PPO/litre and 0·5 ml water. Sources: ^3H–inulin, 222000 dpm, 1·7 µg with no added carrier; ^3H–n–hexadecane, 132000 dpm. Initial counting efficiency 20%. Counting time, 4 min. Instrument set for maximum integral count

ing. This may cause progressive changes in the count rate of both internal and external standards. These artefacts are discussed in more detail in Chapter 11. Also, as mentioned above, the transmission of light through the walls of plastic vials is less than through those of glass ones.

Position and dimensions of counting vial

The position of a vial in relation to the photomultiplier tubes may vary in any of the three dimensions and also rotationally. Its dimensions and the thickness of its wall or base may also vary. Variations of this sort are usually unimportant in the measurement of scintillations derived from a radioactive source within a vial, but they may cause a relatively large error in counting scintillations derived from an external standard.

COUNTING EFFICIENCY AND CHOICE OF RADIOISOTOPE

A number of the factors described above, that is, quenching, self-absorption, instrument instability, variation in volume of vial contents and temperature have their greatest effect when counting efficiency is low. It is therefore advisable, whenever possible, to keep efficiency high by reducing quenching to a minimum and by choosing a radioisotope which can be counted with high efficiency. When there is a choice between ^{14}C and ^{3}H, for example, ^{14}C is unequivocally to be preferred.

For further discussion of the topics mentioned in this chapter, *see* References 58, 83, 94, 104.

CHAPTER FIVE

Selection and quantity of tracer

Before an experiment using a tracer can be carried out, the type of radioactive material must be chosen and the amount required has to be calculated. The final details will obviously depend on the context of the particular experiment, but the basic approach, as discussed in this chapter, is the same in most cases.

CHOICE OF RADIOISOTOPE

A large number of radiochemicals are now commercially available, and for a given compound there is often a choice not only of radioisotope, but also of the site in the molecule at which it is located. The selection of the chemical compound itself is determined in the usual way by the nature of an investigation; the choice of the radioisotope and its position in the molecule may also be limited by the particular requirements of the investigation, such as whether the fate of a specific atom in a compound is to be followed. Whenever possible, the choice should be a long-lived radioisotope which is in a chemically stable site and can be measured with high counting efficiency.

The most common choice is between ^{14}C and ^{3}H, and in this case ^{14}C is preferable because ^{3}H has a number of disadvantages, as follows:

1. Low β–particle energy

The low energy of the β–particles emitted by ^{3}H always results in an efficiency of counting which is lower than that of ^{14}C, and is readily reduced even further by quenching (Fig. 4.6); also this lower efficiency increases the effect of instrument instability on the accuracy of counting (Fig. 4.2). (This disadvantage is not offset by the higher specific activity of ^{3}H compounds which can be achieved.)

2. Higher rate of decay

This may be a disadvantage for two reasons:

1. *Effect on storage.* ^{3}H–labelled compounds usually deteriorate more rapidly during storage (*see* Chapter 2).
2. *Measurable fall in activity.* The rate of decay of ^{3}H, unlike that of ^{14}C, is high enough to result in a significant loss of activity over a period of a few weeks, although in practice a simple correction is all that is required to allow for this (Table 2.6).

3. Relatively large difference between isotopic masses

^{3}H has an atomic mass which is greater by a factor of three than that of the stable isotope ^{1}H. This difference in mass may be sufficient to affect chemical reactions (*isotope effect*)[9,25a,32], and in fact the substitution of ^{3}H for ^{1}H has been reported to reduce the velocity of some reactions by a factor of between 6 and 20[112]. It should therefore not be assumed without evidence that the behaviour of a labelled compound, particularly one labelled with ^{3}H, is identical with that of unlabelled carrier. Although this is the assumption upon which tracer experiments are based, it is seldom checked.

4. The possibility of ^{3}H exchange

In aqueous solutions of labelled molecules, ^{3}H is more likely to be lost by exchange[105] than is ^{14}C.

QUANTITY OF RADIOISOTOPE REQUIRED

In any investigation in which radioactive material is used, it is necessary to know the amount required both of radioactive material (tracer) and of non-radioactive material of the same chemical nature (carrier). The carrier is the compound whose behaviour the investigator wishes to study, and the amount and purity required are determined entirely by the nature of the investigation. Radioactive material is usually supplied mixed with carrier for chemical stability, but the amount and purity of this may be ignored if, as is usually the case, it is subsequently added to a much larger amount of carrier.

The mass of radioactive material used (as distinct from its activity) is usually so much smaller than that of carrier that this may also be ignored. If, however, the radioactive material forms a significant proportion of the whole (that is, of tracer plus carrier), its mass must be taken into account. This may be calculated from the specific activity stated by the supplier; due allowance must be made for decay in the case of a short-lived radioisotope.

The tracer should be of the highest available radiochemical purity, since it is often the sole indicator of the behaviour of the compound under investigation; the presence of even a small proportion of contaminating radioactive material may thus lead to false interpretation.

The amount of radioactive material required in terms of activity depends on the dilution expected, on the time available for counting, on the efficiency of counting and on the accuracy required. Increasing the amount of radioactive material shortens the counting time needed for a given count or increases the count obtained in a given time. However, the amount of radioactive material that may be used is limited by:

1. The possibility of causing tissue damage to personnel or, in biological experiments, to tissues exposed to radiation.
2. The problem of disposal of large quantities of long-lived radio-isotope and the risk of contamination of the environment.
3. The high cost of radioactive material.

In practice, the amount used is a compromise between cost and convenience.

To show how the smallest suitable quantity of radioactive material may be determined, we shall work through an example stage by stage. The simplest method is to start with the count rate eventually required in each sample. The factors to be considered are set out below, and followed by a calculation.

Total count required in each vial

This depends on the accuracy which is required. The statistics of counting are discussed in Chapter 9; briefly, the greater the total count the smaller the error. A minimum total count of not less than 10 000 counts per vial is commonly recommended; this gives a percentage standard deviation of 1% (p. 131), that is, just over 95% of values are within 2% of the mean. If about 95% of values are to be within 1% of the mean (percentage standard deviation 0·5%) a total count of 40 000 counts per vial is needed.

Count rate required

This depends primarily on the time available or the maximum time considered suitable; the shorter the time, the higher the count rate must be to provide the total count required for each sample. A very low count rate should be avoided if possible, since the longer the counting time needed, the greater the error introduced by (i) the contribution of the background count and (ii) instrument instability, particularly at low counting efficiency (*see* Fig. 4.2). Time should be allowed for counting each vial twice as a check for non-random variation.

Counting efficiency

A knowledge of this allows the disintegration rate or activity to be determined from the required count rate. An approximate or estimated value is sufficient for most purposes.

Activity in sample added to vial

The activity in each vial is obtained from a precise quantity of added radioactive sample. The quantity of the sample depends on the amount available, and on its solubility in scintillation mixture. Its activity depends on the amount of dilution (or equivalent) of original material that is expected by chemical reaction, biological metabolism, dispersion or loss.

Example

Number of vials to be counted	10
Total time available for counting	200 min
Total time available for each vial	$\frac{200}{10} = 20$ min
Total count per vial giving 0·5% standard deviation	40 000 counts
Count rate required	$\frac{40\,000}{20} = 2\,000$ cpm
Efficiency of counting	50%
Activity required per vial	$2\,000 \times \frac{100}{50} = 4\,000$ dpm
Volume of radioactive sample added to vial	0·5 ml

Taken from diluted sample of   3 ml

Total activity required in 3 ml sample $4000 \times \dfrac{3}{0.5} = 24000$ dpm

Dilution during experiment   20 x
Activity required before dilution $24\,000 \times 20 = 480\,000$ dpm
Total activity required in 10 samples before dilution
$$480\,000 \times 10 = 4\,800\,000 \text{ dpm}$$
$$\sim 2 \ \mu\text{Ci}$$

The calculated amount of radioactive material is added to sufficient carrier to give the required total mass of the material under investigation. This material then has a known specific activity, that is, its disintegration rate is directly related to its mass. Any subsequent determination of disintegration rate, derived from a measured count rate, can therefore be used to calculate the mass of the material present, or the mass of its derivatives.

In the above example, of course, it is assumed that all samples are diluted to the same extent and that the radioactivity in each vial is the same. In practice this is seldom so, and a compromise must usually be reached in which the amount of radioactive material to be used is calculated from the minimum acceptable count rate of those samples which are expected to contain the lowest amount of radioactivity.

For further discussion of the topics mentioned in this chapter, *see* Reference 104.

CHAPTER SIX

Preparation of sample for counting

In this chapter we shall discuss the constituents of some scintillation mixtures used for counting soluble radioactive materials; special treatments for other radioactive materials will also be described.

SCINTILLATION MIXTURES

A large number of recipes for scintillation mixtures (sometimes known as *scintillation cocktails*) have been published, but for many purposes a few simple ones will suffice[56]. The properties of a suitable mixture are as follows:

1. It should generally be clear, colourless and uniform after addition of the radioactive sample, although in certain cases suspensions or gels may be satisfactory.
2. It should quench as little as possible. This is particularly important when measuring radioisotopes, such as ^{3}H whose β–particles are of low energy.
3. It should not be expensive. Only in special circumstances, such as when solubilisers are used (*see* p. 75), are costly mixtures and 'scintillation grade' chemicals of high purity a significant improvement on simpler and cheaper ones[107].
4. Its constituents should be stable. Some scintillators are known to be unstable when exposed to light[1], and other materials may deteriorate on storage, with the formation of impurities which

cause quenching or chemiluminescence. Scintillation mixtures and their constituents should therefore be stored in dark bottles.

Because of possible variations in purity and stability, the scintillation mixture used in any one set of measurements should always be taken from the same stock solution.

Mixtures suitable for toluene-soluble samples

The simplest mixture consists of a primary scintillator (and, if necessary, a secondary scintillator) dissolved in a primary solvent. Primary solvents are usually aromatic hydrocarbons, and therefore only non-polar radioactive materials can be dissolved in them. For aqueous solutions more complex mixtures must be used, and these will be described later.

The most widely used primary solvent is toluene. Other primary solvents are $p-$, $m-$ and mixed xylene, with which the efficiency of counting is up to 10% greater than with toluene[12]. Many more have been used, but are more expensive, and show no marked advantage over toluene.

The cheapest laboratory grade of primary solvent is often as satisfactory as the most expensive; we have found less than 1% difference in the efficiency of counting ^{3}H$-n-$hexadecane in a rectified and in a scintillation grade toluene from a commercial source, while there was no significant difference in background count rate. However, a scintillation grade solvent is essential in certain circumstances; for example, inferior grades of toluene may cause excessive chemiluminescence with solubilisers (p. 136).

Two commonly used primary scintillators are *PPO* (2, 5—diphenyl-oxazole) and *butyl—PBD* (2—(4'—*tert*—butylphenyl)—5—(4"—biphenyl)—1, 3, 4—oxadiazole). PPO may be slightly the cheaper of the two in use, but the pulse voltage produced by butyl—PBD is about 20% greater than that produced by PPO. In theory, therefore, butyl—PBD should be the better scintillator at low counting efficiency, although this is not always the case, as can be seen in Table 6.1. (The parent compound of butyl—PBD, PBD, produces a greater pulse voltage, but limited solubility reduces its usefulness.) If a solubiliser is to be used in the presence of butyl—PBD its effect should be checked; it has been reported that severe colour quenching may be caused by the interaction of butyl—PBD with some commercial solubilisers[30, 63] (*see* Table 6.3).

Suitable concentrations of PPO and butyl—PBD are about 5 g/l and 7 g/l respectively. In highly quenched samples, a higher concentration may be required for optimum efficiency, as shown, for example, in Fig. 6.1, but unless efficiency is very low the extra cost does not usually

TABLE 6.1. Efficiency of counting in scintillation mixtures described in the text

About 10 mg ^{14}C–n–hexadecane (2978 dpm/mg) or ^{3}H–n–hexadecane (4700 dpm/mg) were dissolved in 10 ml scintillation mixture. No secondary scintillator was used in Mixtures 1–4. Values are the mean of duplicate samples. Counting time 1 min. Instrument set for maximum integral count. Figures in heavy type represent efficiencies after the addition of 1 ml of water

Mixture no.	Constituents	Efficiency of counting, %			
		^{14}C		^{3}H	
		PPO	Butyl–PBD	PPO	Butyl–PBD
1	Toluene	94 –	97 –	55 –	59 –
2	Toluene/cellosolve (1:1)	87 **87**	90 **86**	24 **19**	32 **25**
3	Toluene/'Triton X-100' (2:1)	92 **89**	93 **87**	41 **33**	41 **32**
4	Toluene/'Triton X-100' (1:1)	90 **89**	89 **88**	36 **28**	30 **22**
5	Bray's Fluid	**92**	**89**	**35**	**23**

warrant what may be only a marginal improvement. Excessive concentration of primary scintillator reduces efficiency, as shown in the figure.

A secondary scintillator may be needed in the mixture if the emission wavelength of the primary scintillator does not match the wavelength to which the photomultiplier is most sensitive, but it is often unnecessary with photomultipliers now in use. It may still improve performance

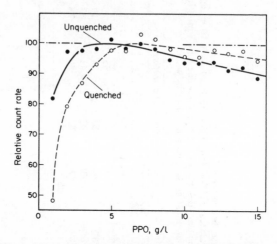

Fig. 6.1. Effect of concentration of PPO on count rate. Identical quantities of
^3H–n–hexadecane were added to glass counting vials each of which contained 10
ml toluene with PPO at the concentration shown. Closed circles *represent*
samples which were unquenched; relative count rate of 100 = 49·4% counting
efficiency (19 590 cpm). Open circles *represent samples which were quenched by*
the addition of 0·5 ml chloroform; relative count rate of 100 = 17·4% counting
efficiency (6 899 cpm). Counting time, 10 min. Instrument set for maximum
integral count

when counting efficiency is low, or when plastic vials are used; the benefit may be marginal, and must be set against the extra trouble and cost. Commonly used secondary scintillators are *POPOP* (1, 4–di– [2– (5–phenyloxazolyl)]–benzene) and *dimethyl–POPOP* (1, 4–di–[2– (4–methyl–5–phenyloxazolyl)] –benzene), used at concentrations varying from 0·05 to 0·5 g/litre (usually 0·1 g/litre). POPOP is the less soluble of the two but is slightly cheaper. Both should be given adequate time to dissolve completely, preferably overnight.

A suitable mixture for toluene-soluble materials is:

MIXTURE 1.
   Toluene containing 5g PPO/litre or 7 g butyl–PBD/litre.
   POPOP or dimethyl–POPOP, 0·1 g/litre, is added if necessary.

Mixtures suitable for aqueous samples

Mixture 1 cannot be used for measuring the radioactivity in aqueous samples since water is immiscible with toluene. Further materials must be added to this mixture (or substituted for toluene) to enable water to be incorporated; unfortunately all such additions cause quenching, as indeed does water itself. Because of this, the lower the energy of β–emissions, the greater the care required in the choice of the constituents of a scintillation mixture; for example, two mixtures in which ^{14}C is counted with approximately equal efficiencies may show markedly different efficiencies in the counting of ^{3}H, as shown in Table 6.1.

Two types of mixture are commonly used, one based on toluene, the other containing 1, 4–dioxane. The first type is usually simpler to prepare, is considerably cheaper and is less subject to deterioration.

1. *Mixtures based on toluene*

These usually consist of a simple mixture such as Mixture 1, to which is added a further solvent (*secondary solvent*) miscible with both water and primary solvent. (Certain types of secondary solvent are sometimes referred to as 'solubilisers', but the latter term will be restricted in this book to its more usual meaning of digesting agents, p. 75.) The amount of aqueous solution which can be incorporated in a particular mixture without phase separation depends on the identity of the secondary solvent, the relative proportions of primary and secondary solvents and the temperature.

It is essential to check that a mixture containing water does not separate into two phases under operating conditions, since the physical separation of a water-soluble labelled compound from the primary solvent and scintillator causes a large reduction in count rate.

As a general rule, the greater the ratio of secondary to primary solvent, the greater the amount of water which can be incorporated, but the extra secondary solvent and the extra water increase the quenching.

Two satisfactory secondary solvents are *cellosolve* (2–ethoxy-ethanol)[44] and *'Triton X–100*[76, 100]. Cellosolve acts by dissolving in both toluene and water. 'Triton X–100', on the other hand, is a surface active agent, and an apparently homogeneous mixture with toluene and water is believed to consist of a suspension in primary solvent of 'Triton X–100' micelles containing water.

The following mixtures are suitable for most types of aqueous solutions; they are simple to prepare and use, give reasonable efficiencies and are relatively cheap.

MIXTURE 2.

Toluene   500 ml ⎫
Cellosolve   500 ml ⎭ + 5 g PPO/litre or 7 g butyl–PBD/litre

POPOP or dimethyl–POPOP, 0·1 g/litre, is added if necessary.

Up to about 1 ml of water mixes homogeneously with 10 ml of mixture.

MIXTURE 3.

Toluene   667 ml ⎫
'Triton X–100'   333 ml ⎭ + 5 g PPO/litre or 7 g butyl–PBD/litre

POPOP or dimethyl–POPOP, 0·1 g/litre, is added if necessary.

Not less than about 0·5 ml and not more than about 1·4 ml of water mixes homogeneously with 10 ml of mixture.

MIXTURE 4.

Toluene   500 ml ⎫
'Triton X–100'   500 ml ⎭ + 5 g PPO/litre or 7 g butyl–PBD/litre

POPOP or dimethyl–POPOP, 0·1 g/litre, is added if necessary.

Up to about 2 ml of water mixes homogeneously with 10 ml of mixture.

The concentrations of primary scintillator given above are those used in Mixture 1, and are suitable for most purposes. It may be found that higher concentrations of scintillator than those shown partly compensate for the quenching caused by secondary solvent and aqueous material. However, the increase in efficiency thus obtained is usually too small to justify the increase in cost. For example, an increase in the concentration of PPO from 5 to 7 g/litre for the quenched sample in Fig. 6.1 increased efficiency by only about 5%, but increased the cost of PPO by about 40%.

The proportions of solvent shown in Mixture 3 are those usually recommended with 'Triton X–100'[76, 100]. We have found that Mixture 4 is sometimes preferable, since water mixes more readily in all proportions and counting is more reproducible, but the efficiency of counting for ^{3}H is less owing to the higher concentration of 'Triton X–100' (Table 6.1).

If the volume of water added to Mixtures 3 and 4 is progressively increased, a milky stage is reached and then a gel is formed which may be used for supporting suspensions or larger volumes of water (*see* below).

2. *Mixture based on 1, 4–dioxane*

1, 4–dioxane, which is miscible with water, may wholly or partly replace toluene, and thus act as a primary solvent. The efficiency of counting is only about 70% of that with toluene, yet it is about six times more expensive. In addition, it may contain peroxides which cause quenching and may contribute towards chemiluminescence. It may thus need to be purified before use, although subsequent deterioration may be inhibited by adding 0·001 % sodium diethyl dithioarbamate or butylated hydroxytoluene[58]. Its freezing point is 12°C, so caution should be used with refrigerated instruments.

We give here the most commonly used mixture, that of Bray[19, 20].

MIXTURE 5.

Methanol (absolute)	100 ml
Ethylene glycol	20 ml
Naphthalene	60 g
PPO	4 g
POPOP	0·2 g
1, 4–dioxane	to make 1 litre

(Naphthalene has been incorporated to reduce quenching[20].)
Up to 3 ml of water mixes homogeneously with 10 ml of mixture.

The sole advantage of this mixture is that it can incorporate more water than many others, but in most respects it is less satisfactory than Mixtures 2 − 4.

OTHER TECHNIQUES IN SAMPLE PREPARATION

Use of gels for counting aqueous samples[56]

Most of the mixtures described above will only take up a relatively small volume of aqueous solution. Each of the two which contain 'Triton X−100', however, can be used in the form of a gel. Progressive increase in the amount of water added to the basic mixture results first in a milky stage, in which counting efficiency falls and becomes erratic, and then a stable gel stage is reached. Such a gel is formed if more than about 2 ml of water are added to 10 ml of Mixture 3[99], while a 1:1 (v/v) mixture of water and Mixture 4 also produces a satisfactory gel. The counting efficiency is, however, relatively low owing to the quenching caused by the large amount of water present.

Increasing the amount of aqueous sample in a gel up to a certain percentage of the total volume raises the count rate; above this percentage the count rate falls because of the increase in quenching. This relationship is sometimes expressed by a figure of merit[56], which is calculated as 'efficiency x percentage aqueous sample'. The highest figure of merit corresponds to the maximum count rate.

Relatively large volumes of water may be counted with high

efficiency by the use of ready-prepared commercial gelling agents, although high cost may preclude their routine use in large quantity. A gelling agent may, however, be used at low cost in the following manner[72a]. Aqueous sample (roughly neutralised) is mixed with a suitable volume of gelling agent in a small glass specimen tube, which is then placed inside a glass counting vial for counting. For example, ^{14}C in 1 ml of water mixed with either 1 ml of 'Instagel' (Packard Instrument Ltd.) or 3 ml of a toluene/'Triton X–100' (2:1 v/v) based mixture could be counted with similar efficiency and reproducibility as in a more orthodox system using 10 ml of scintillation mixture, but the effective cost was less. Specimen tube and contents are discarded after use, yet the advantages of a glass container (p. 155) are retained.

Error in counting aqueous samples with reagents of this type may be caused by differences in the distribution of sample, quenching agents and radioactive standard. In a mixture consisting of water-containing micelles suspended in non-aqueous medium, for example, it is possible for a water-soluble radioactive material to be counted at different efficiency from a toluene-soluble radioactive standard, for the one is in solution in the aqueous phase and the other in the non-aqueous phase; the standard cannot then be used for determining the efficiency of counting of the water-soluble material without confirming that they are counted with the same efficiency. This can be done by checking that the channels ratio (p. 99) of the aqueous sample is the same as that of a toluene-soluble standard counted under exactly the same conditions.

Coloured solutions

Coloured materials present in the counting vial may absorb some of the light from scintillations (colour quenching). This effect is naturally greatest when the sample absorbs most strongly those wavelengths emitted by the scintillator. For example, the peak emission wavelength of PPO is about 3 700 Å, and it is therefore affected most by red solutions and least by blue solutions. The light emitted by secondary scintillators has a longer wavelength (e.g. dimethyl–POPOP, 4 300 Å) than that emitted by primary scintillators, and this shift of wavelength may therefore reduce some types of colour quenching. Light may similarly be absorbed by polyethylene vials, and to some extent this may also be overcome by the addition of secondary scintillator.

A quench correction curve (Chapter 7) prepared with a chemical quenching agent cannot be used to correct for colour quenching if the range of quenching of the samples is large[21, 74]. If possible, therefore, excessive colour should be removed before counting.

Colour may often be removed by a bleaching agent, which itself, however, may cause chemical quenching. Probably the simplest and cheapest methods for biological samples are those in which hydrogen peroxide[50,66], or benzoyl peroxide[45,103] are used, but there may be disadvantages with these materials[10] and hydrogen peroxide gives rise to chemiluminescence (p. 136) with solubilisers[110] (as indeed do other peroxides). Chlorine water has also been used[96].

It is worth remembering that reducing the volume of sample added to a given volume of scintillation mixture will reduce the concentration of coloured material and hence will raise the efficiency of counting.

A combustion technique may provide a suitable alternative for the treatment of coloured samples.

Combustion techniques

The combustion of a sample in an atmosphere of oxygen may sometimes be a more satisfactory way of preparing radioactive material for counting than either decolorisation or solubilisation[34]. An advantage of combustion is that it converts any given radioisotope into one chemical form no matter what the nature of the original material; thus all ^{3}H– and ^{14}C–labelled compounds are counted as ^3H_2O and $^{14}CO_2$ respectively. Recoveries are said to be good[40]. It is a technique which is particularly suitable for the counting of samples containing mixed ^{3}H and ^{14}C, for these radioisotopes may then be almost completely separated from each other, so that mutual interference in counting is greatly reduced.

It would be out of place to discuss methods in detail here, and the reader is referred to other sources[58,60,75,82,100]. Various types of automatic combustion apparatus are commercially available[84b]. Other oxidation methods have been used[24].

Biological material[56, 58, 79, 81, 85, 87, 100]

It is often possible to convert organic compounds, such as protein, into a more soluble form by means of a suitable *digesting agent*, more commonly but inelegantly known as a *solubiliser*. (Occasionally this term is used to describe a type of secondary solvent.) This is a strongly basic liquid, and may have several disadvantages, such as the production of chemiluminescence (*see* p. 136 and Table 6.2) and quenching (Table 6.3).

Some common solubilisers are: 'Hyamine 10–X hydroxide', which is still frequently used in spite of its disadvantages, 'NCS Solubilizer' (Nuclear-Chicago Corporation), 'Soluene' (Packard Instrument Ltd.)

TABLE 6.2. Chemiluminescence caused by various reagents in the presence of solubiliser

The reagents were added in series as follows: 5 ml toluene containing 5 g PPO/litre and, where stated, 0·3 ml solubiliser were mixed in a glass counting vial and counted immediately; after the count rate had stabilised in the dark, 5 ml cellosolve containing 5 g PPO/litre was added and the mixture counted again; after stabilisation of the count rate again, 0·1 ml hydrogen peroxide was added and the mixture counted. Samples were prepared in dim light. In each case the count was taken at the times shown after mixing, and the values shown are the approximate count rates in cpm after subtraction of background count rate. Toluene was of scintillation grade. Counting time, 1 min. Instrument set for maximum integral count

Solubiliser	Toluene reagent + solubiliser	Toluene reagent + solubiliser + cellosolve reagent			Toluene reagent + solubiliser + cellosolve reagent + hydrogen peroxide	
	1 min	1 min	10 min	60 min	1 min	10 min
No solubiliser	nil	10	10	35	35	10
'Hyamine 10–X Hydroxide'*	nil	590	255	5	25 505	60
'NCS Solubilizer'**	35	1780	340	35	33 290	75
'Protosol'φ	120	1675	405	15	40 815	145
'Soluene'*	100	1730	200	40	47 035	75

* Packard Instrument Ltd.
** Nuclear-Chicago Ltd.
φ New England Nuclear

TABLE 6.3. Effect of some solubilisers on efficiency of counting
Radioisotope was added to 10 ml toluene/cellosolve (1:1 v/v) containing 5 g PPO/
litre or 7 g butyl–PBD/litre and either 1 ml toluene or 1 ml solubiliser. Samples
were counted after 48 hr. The count rates of samples with solubiliser are compared
below with those of samples without solubiliser and expressed as a reduction in
count rate. Sources: ^{14}C–n–hexadecane, 2980 dpm; ^{3}H–n–hexadecane, 4 700
dpm. Counting efficiency in absence of solubiliser: ^{14}C, 90%; ^{3}H, 32%. Counting
time, 10 min. Instrument set for maximum integral count

Solubiliser	Reduction in count rate (%)		
	PPO		Butyl–PBD
	^{14}C	^{3}H	^{3}H
'Protosol'**	4	19	33*
'Soluene'$^{\phi}$	5	20	86*
'NCS Solubilizer'$^{\phi\phi}$	7	22	38*
'Hyamine 10–X hydroxide'$^{\phi}$	10	40	52

* These samples turned yellow, causing progressive colour quenching
** New England Nuclear
ϕ Packard Instrument Ltd.
$\phi\phi$ Nuclear-Chicago

and 'Protosol' (New England Nuclear). The method of use is similar for
all of them. Usually a quantity of solubiliser, say 0·5 to 1·0 ml, is mixed
in a counting vial with the material to be dissolved, and left to stand
(warmed if necessary) for several hours. Scintillation mixture is then
added in the usual amount, and the vial counted after any chemiluminescence has disappeared or been eliminated. The use of these materials
may lead to coloured products, with consequent colour quenching; for
example, this may occur with certain solubilisers when a scintillation
mixture containing butyl–PBD is used[30, 63] (*see* Table 6.3) or if they
are warmed above about 60°C.

Inorganic salts[70]

Salts which will not remain in solution in scintillation mixture may be
counted after suspension in a gel (*see* below). Alternatively, a salt
containing, say, ^{45}Ca or ^{55}Fe may be converted to a toluene-soluble
complex[27, 46] which is then dissolved in Mixture 1.

Suspension of insoluble materials in gels[39, 82, 94, 100]

Insoluble particles cannot be counted satisfactorily in a liquid scintillation mixture owing to the progressive fall in count rate that occurs as

the particles settle; instead, they may be suspended in a rigid or semi-rigid matrix such as a gel. Particles should be as fine and as uniform as possible, since self-absorption occurs with solid material, and the quenching which results not only reduces the overall efficiency of counting, but may vary from sample to sample, depending on the uniformity of particle size. These faults are particularly noticeable with ^{3}H because of its marked susceptibility to quenching.

Several materials, of which the following are examples, have been found suitable for preparing suspensions:

1. Gels based on silica (e.g. 'Cab-O-Sil', Packard Instrument Ltd.; 'NE 221', Nuclear Enterprises Ltd.). The combination of a suitable scintillation mixture and powdered silica is thixotropic, that is, it forms a gel at rest but liquefies on shaking.
2. Other types of gel. A toluene/'Triton X−100' mixture such as Mixture 3 or 4 forms a gel on the addition of a suitable amount of water. Some commercial scintillator-gelling agent mixtures, such as 'Insta-gel' (Packard Instrument Ltd.) have similar properties. The quantity of water to be added to obtain a suitable gel is fairly critical for best results; a nearly equal volume of water should be added to Mixture 4 or to 'Insta-gel', and somewhat less to Mixture 3. The mixture is rigid and transparent at about $5°C$ but is semi-rigid and opalescent at about $20°C$. We have found that the reproducibility of counting ^{14}C and ^{3}H at $5°C$ with Mixture 4 prepared in this way is slightly less than in a homogeneous liquid mixture; a gel derived from Mixture 3 has been reported to be unreliable for counting ^{3}H[99].

As an alternative to the suspension of insoluble materials, combustion may sometimes prove satisfactory.

Material in polyacrylamide gels[41,69]

After separation of radioactive materials by the technique of electrophoresis in polyacrylamide gels, the radioactivity of the separated components in the gel can be determined in several ways. The simplest is by elution of material from the gel, although recoveries by this method may vary[113]. This may be carried out as follows:

1. Slice the gel into the appropriate number of sections, depending on the resolution required.
2. Place each piece into a small test tube, add a little sand or ground glass and break up the piece of gel with a glass rod.

3. Add water, buffer, etc. as required.
4. Centrifuge when extraction is complete.
5. Take the required quantity of supernatant for counting.

Some methods make use of solubiliser for extraction[4]; in others the gel may be disintegrated by peroxide[37, 98], or special gels which dissolve in solubiliser may be prepared[77]. Combustion methods may also be used.

Material on solid supports[28, 35, 58, 82]

Radioactive material deposited on a solid support, such as glass fibre discs, chromatography paper, filter paper or other filtration media, may be measured by placing the whole in a glass vial and adding a simple scintillation mixture such as Mixture 1.

The efficiency of counting may be low and reproducibility poor, especially with ^3H, owing to absorption of the energy from β–particles by the material of the support. For example, with filter paper as support, the efficiency of counting has been reported to be not more than 60 or 70% for ^{14}C and only about 5% for ^3H[17].

In a similar way, radioactivity on thin layer chromatograms may be determined by scraping off the surface material into vials.

One of the disadvantages of counting radioactive material on solid supports or as a suspension is the difficulty of determining counting efficiency (Chapter 7). External standards[28, 35] and toluene-soluble internal standards are unsuitable, because they measure the efficiency of counting in the solution only, and take no account of quenching caused by the support or by the insoluble material itself. The sample channels ratio method is the only suitable method of determining efficiency, and even this is difficult to use accurately in these circumstances (*see* p. 113).

Efficiency may often be raised and more reliably determined by elution of the material from the solid support in the vial[65, 97]. Each piece of paper or other material on which the sample is deposited is placed in a separate vial, and a small amount of a suitable solvent is added. This is followed when elution is complete by scintillation mixture. Silica gel from thin layer chromatograms may alternatively be dissolved in hydrofluoric acid before counting[95].

Counting of gases and vapours

Gases and vapours may be counted either by first trapping in a suitable solution or by incorporating directly in a liquid scintillation mixture.

As a simple example, carbon dioxide may be trapped in an alkaline solution before being counted. There are many possible ways of preparing gaseous materials for counting, and for more detail we refer the reader to other sources[13, 58, 100, 102, 109].

ČERENKOV COUNTING[82, 100]

β—particles of sufficiently high energy cause the emission of light in aqueous solution without the addition of primary solvent or scintillator. This light is called Čerenkov radiation and is said to be due to an electromagnetic shock wave effect[36]. It is detectable in a liquid scintillation counter and satisfactory counting efficiencies are obtained with β—particle energies greater than about 1 MeV[100]. Thus ^{32}P (1·71 max. MeV), for example, may be counted in this way[47].

The aqueous solution is simply placed in a vial and counted, and the sample thus remains unaltered; at the same time, chemical quenching does not occur. Colour quenching, however, may be present as in normal liquid scintillation counting.

COUNTING OF γ—RAYS

γ—rays may be counted in a liquid scintillation counter without interference from α— or β—particles. A special type of vial is used in which there is a central glass well surrounded by scintillation mixture containing a γ—ray absorbing material such as tetrabutyltin sealed off from the exterior[2a]. The γ—ray emitter is placed in a container in the central well and the resulting scintillations counted in the usual way. The problem of miscibility of the sample with scintillation mixture is avoided, and, as far as we know, there is no chemical or colour quenching. Commercial vials are obtainable, but are expensive. In using this technique it is important to assess interference by γ—rays arising from the contents of neighbouring vials.

DETERMINATION OF IDENTITY OF LABELLED COMPOUNDS

A check should always be made that the total activity in the various derivatives of the original material is equal to the amount of activity originally present, that is, a complete 'balance sheet' of the radioactive materials in an experiment should be prepared. This helps to guard not only against unawareness of significant losses, but also against the ever-present danger of contamination by extraneous radioactive material; if any radioactive contaminant enters experimental materials from, say, inadequately washed apparatus, this should be indicated by a greater total activity at the end of the experiment than was introduced as tracer.

It is also important to remember that the measurement of radioactivity indicates how many tracer atoms are present, but does not identify their chemical form. The count recorded could be due to unchanged original material, to derivatives or even to contamination, possibly by a different radioisotope.

The identity of derivatives and the amount of activity in each must be determined after they have been separated by suitable means. The purity of the originally supplied radioactive compound should also be checked before use, particularly if it has been stored for any length of time. One of the simplest analytical methods is paper (or thin layer) chromatography.

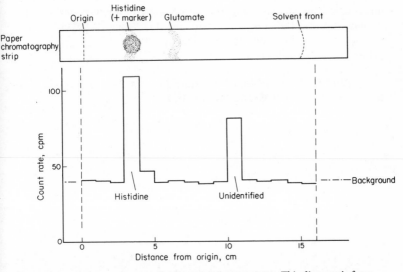

Fig. 6.2. Paper chromatogram of radioactive tissue extract. This diagram is from an experiment whose purpose was to investigate the metabolism of histidine in brain slices. Slices were incubated for 1 min in Krebs' bicarbonate saline which contained 0·005 mM L-histidine incorporating 0·01 μCi/ml ^{14}C–L–histidine, and then extracted by homogenisation in 5% trichloroacetic acid. Extract was applied to a paper chromatogram strip; extract containing added unlabelled histidine as marker was applied to a second strip for locating amino acids by colour development with 0·2% ninhydrin in acetone. After running in butanol:acetic acid:water (120:30:50) the first strip was cut into 1 cm segments, and each segment placed in a glass counting vial together with 10 ml toluene containing 5 g PPO/litre. Background count was obtained from extract under identical conditions but with no added radioactive material. Counting time per vial, 40 min. Instrument set for maximum differential count of ^{14}C

Paper and thin layer chromatograms[23, 100]

A sample of the material to be examined is applied to a strip of
chromatography paper or a thin layer plate and run in a suitable solvent.
The paper is then cut transversely into, say, 1 cm strips, and each strip
is placed in a vial with 10 ml of Mixture 1. In a similar way, sections of
a thin layer chromatogram are scraped off the plate and put into vials
with scintillation mixture. The vials are then counted, and the count
rate from each is plotted graphically against its distance from the
origin, as shown in Fig. 6.2. The identity of the peaks of radioactivity
must be established by the usual methods. A strip from a chromatogram
run using only blank materials must also be counted to provide a base
line.

The efficiency of counting may be low, and the quantitative determination of radioactivity on paper chromatograms presents special
problems; it may therefore be advisable to elute from the paper in the
vial (*see above* under 'counting on solid supports').

For further discussion of topics mentioned in this chapter *see*
References 56, 58, 80, 81, 82, 87, 88, 100.

CHAPTER SEVEN

Determination of efficiency (quench correction)

On first acquaintance with liquid scintillation counting it is easy to make the serious mistake of assuming that no more is required for the determination of radioactivity than to dissolve the sample in a suitable scintillation mixture and then to count it. This would be possible only if a given amount of a radioisotope gave the same count rate whatever the conditions under which it was counted, but this is unfortunately not the case. The count rate is *not* a direct measurement of the amount of radioactivity and cannot be used to determine it without a knowledge of the efficiency of counting.

The count rate alone thus cannot be used for the valid comparison of samples, either with each other or with standards, unless all samples are known to be counted with exactly the same efficiency. However, in practice efficiency frequently varies. Even when all the controllable variables associated with counting are eliminated, quenching caused by material introduced with a sample (or even caused by the radioactive material itself) frequently differs from one sample to another. For example, Fig. 7.1 shows the uncorrected results of counting a strongly quenching biological material. Although the concentration of radioactive material was the same in all samples, the count rate was markedly affected by changes in the concentration of haemoglobin. This is an extreme example, but it is clear that large errors of interpretation may occur unless quenching is taken into account.

Samples can only be compared quantitatively, therefore, by determining the efficiency of counting of each sample and using this to

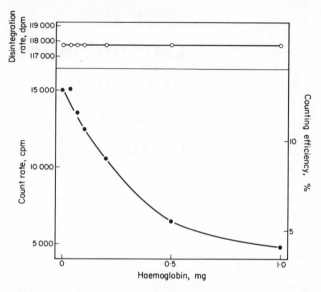

Fig. 7.1. Effect of a coloured protein (haemoglobin) on count rate and efficiency. Counting vials containing the desired quantity of haemoglobin in 0·2 ml water were mixed with 0·3 ml 'Hyamine 10–X hydroxide' and incubated at 50°C for 70 min. 10 ml scintillation mixture (toluene/cellosolve, 1:1 v/v), containing 4 g PPO/litre with 25 mg ³H–n–hexadecane (4710 dpm/mg) was added. Counting time, 1 min. Instrument set for maximum integral count. It may be noted that haemoglobin causes both chemical quenching and colour quenching

convert its count rate to a disintegration rate; comparisons are otherwise meaningless.

We shall give five methods for finding the efficiency of counting of samples with varied quenching. It is first necessary, however, to discuss the way in which quenching affects pulse voltages in order to understand certain of these methods, and to describe the calibration standards which are often used as a basis for the determination of efficiency.

EFFECT OF QUENCHING ON THE PULSE VOLTAGE SPECTRUM

Any form of quenching causes a reduction in the intensity of the light reaching the photomultiplier tube. Since the pulse generated by a photomultiplier tube and amplifying circuit is proportional to the light intensity, the voltage of all pulses is reduced by quenching. The pulse voltage spectrum as a whole is therefore shifted to a lower voltage range,

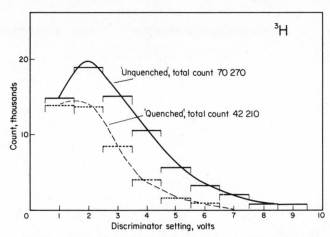

Fig. 7.2. Effect of quenching on pulse voltage spectrum of ³H. Radioactive material was counted in a mixture of 20 μl 0·14 M NaCl, 1 ml cellosolve and 9 ml toluene containing 5 g PPO/litre; a further 0·2 ml toluene (to give 'unquenched' sample) or 0·2 ml chloroform as quenching agent (to give 'quenched' sample) was added. Spectra obtained as described in legend to Fig. 3.6, using the balance point setting of 'unquenched' sample counted in the window 0·5–9·5 V. Counting efficiency over range 0·5–9·5 V: 'unquenched', 43%; 'quenched', 26%.
Source: ³H–n–hexadecane, 163 050 dpm. Counting time, 1 min

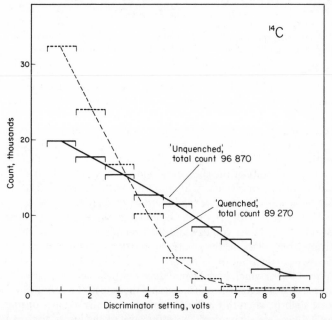

Fig. 7.3. Effect of quenching on pulse voltage spectrum of ¹⁴C. Details as in legend to Fig. 7.2. Counting efficiency over range 0·5–9·5 V: 'unquenched', 88%; 'quenched', 81%. Source: ¹⁴C–n–hexadecane, 110 090 dpm

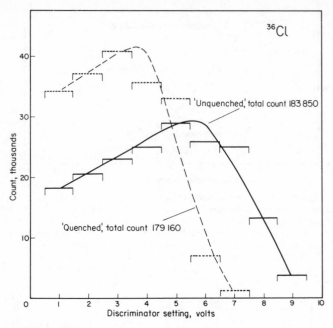

Fig. 7.4. Effect of quenching on pulse voltage spectrum of ^{36}Cl. Details as in legend to Fig. 7.2. Source: Na^{36}Cl (specific activity not known)

and the total number of pulses generated is usually reduced because of an increase in the number of scintillations which are too faint to be detected.

The effect of quenching on the pulse voltage spectra of radioisotopes with low, medium and high energies of emission (^{3}H, ^{14}C and ^{36}Cl respectively) are shown in Figs. 7.2–7.4. It can be seen that quenching resulted in (i) a shift of the spectrum to lower voltages and (ii) a fall in overall count rate or efficiency of counting. With ^{14}C and ^{36}Cl the count rate at the lower voltages rose, and there was a relatively small loss in overall efficiency. With ^{3}H, on the other hand, there was no increase in count rate at the lower voltages and the overall loss of counts was relatively large, owing to the greater proportion of scintillations originally near the limit of detection.

Pulse voltages which have been reduced in this way can be increased by decreasing the attenuation (increasing the gain). The count rate may increase slightly when this is done, depending on the number of pulses whose voltages are then raised above the level of the lower discriminator; counts 'lost' as a result of the detection of fewer scintillations by the

photomultiplier tube cannot, however, be regained. It also follows that
the balance point for a quenched sample is at a lower attenuator setting
(higher gain) than that for an unquenched sample of the same radioisotope.

RADIOACTIVE STANDARDS

A radioactive standard of accurately known activity is essential for the
determination of the efficiency of counting. The use of radioactive
standards is in principle similar to that of standards in colorimetric or
other types of chemical analysis, but the unit of reference is disintegration rate instead of mass.

There are two types of radioactive standard: (i) an *internal standard*,
which is a β-emitter of accurately known activity used for calibration
purposes, and which, if dissolved in unquenched scintillation mixture,
provides a *reference standard*; (ii) an *external standard*, which is a γ-
emitter incorporated into most instruments.

Reference standard

This may be made in a laboratory from a suitable β-emitter, but a
sealed reference standard, of the type supplied with most liquid scintillation counters, is usually preferable, since it is prepared in such a way
as to minimise deterioration. It commonly consists of scintillator
dissolved in primary solvent containing a small proportion of ^{14}C- or
^{3}H-toluene in a permanently sealed vial, with the date of preparation
and the activity at that time marked on the label. It should be used
only to check instrument performance, and can give no indication of
the efficiency with which experimental samples (of different composition and possible variation in quenching) are counted. It is nevertheless advisable to count a reference standard with each batch of samples,
since any unaccountable gross variations in count rate can then be
classified immediately either as a fault in the instrument (and this is not
as uncommon as manufacturers would have us believe) or as a fault in
sample preparation.

Internal standard

This type of standard can be used to determine counting efficiency
either by adding it to the vial containing the sample (direct method) or
by using it to prepare a separate calibration curve (indirect methods).
Analytical details provided by the supplier of such a standard should
state its precise specific activity on a given date, and the efficiency of

counting the standard may then be obtained from the count rate of an accurately measured quantity.

An internal standard as usually used in a liquid scintillation mixture should fulfil certain criteria. It should:

1. Be readily soluble in scintillation mixture.
2. Not itself cause quenching in the amounts normally used.
3. Be of high specific activity.
4. Be chemically stable.
5. Be in liquid form to enable small amounts to be dispensed easily.

These criteria are generally fulfilled by toluene or n-hexadecane labelled with ^{14}C or ^{3}H, which are the internal standards now commonly available. n-Hexadecane is usually preferable, since toluene is readily lost by evaporation. It should not be forgotten that the disintegration rate of a ^{3}H standard decreases by about 0·5% in a month (Table 2.6).

Other radioactive materials, such as those used in experiments, are not usually satisfactory standards, partly because the specific activity is not known with sufficient accuracy and partly because the radioactive material itself may sometimes cause quenching. However, a non-quenching radiochemical can be used as an internal standard in special circumstances if its exact specific activity is first determined by comparison with a commercial standard.

An internal standard should be dispensed by weight rather than by volume, since weight is not temperature-dependent and the error of measurement by weight is usually smaller. We have found that when dispensing 10 μl (a quantity commonly used) by means of a semi-automatic pipette the standard deviation of the count rate was about 2%, whereas when weighing 10 mg on a balance accurate to 0·1 mg it was less than 1%.

An internal standard may easily be dispensed for weighing by drawing it into a short length of narrow-bore polythene tubing fitted over a needle attached to a 1 ml syringe and then expelling it dropwise. (The polythene tubing must always be discarded after use.)

An internal standard may be used either directly or indirectly to measure efficiency. In the *direct method* it is added to the vial containing the sample whose activity is to be found, and the efficiency is determined from the resulting additional count rate. In *indirect methods* a separate series of quenched standards is prepared and used to construct an efficiency calibration curve, more usually known as a *quench correction curve*, which relates the counting efficiency of each of the standards to some measurement of quenching. A corresponding measurement is obtained from each sample of unknown activity and the efficiency of counting obtained from the curve.

Quenched standards

These consist of a series of vials which contain a known amount of internal standard together with graded quantities of a quenching agent for the indirect determination of counting efficiency. They thus provide a series of standards whose disintegration rates are known and whose differences in counting efficiency are caused by the quenching agent.

Vials are prepared which contain the following:

1. Scintillation mixture and any other non-radioactive materials which are present in the vials containing the sample whose efficiency of counting is to be determined.
2. Accurately measured amounts of an internal standard labelled with the same radioisotope as the samples.
3. A quenching agent, such as chloroform, in graded amounts from zero upwards, such that the counting efficiency of the most highly quenched standard is at least as low as that of the most quenched sample.

The vial which contains no added quenching agent is usually known as the *least quenched standard*. The total volume in each vial must be the same (*see* Fig. 4.4), and any difference should be made up with primary solvent. An example of a series of quenched standards for ^3H and ^{14}C is shown in Table 7.1. When a *blank* is required, it normally has the same composition as the least quenched standard, but without radioactive material.

It is often assumed that a single quench correction curve for each radioisotope can be used to determine the reduction in efficiency caused by a wide range of quenching agents, and commercially prepared *sealed quenched standards* are therefore often used in place of quenched standards prepared in the laboratory. It is, however, bad experimental practice in any method of assay to use standards consisting of materials which are known to be different from those of the samples with which they are to be compared. Ideally, of course, the quenching material whose concentration may vary in the experimental samples should be used as the quenching agent in the standards. This is particularly desirable with samples which are coloured, since it has been reported[3, 21, 74] that the effect of colour quenching is not always adequately corrected for by a quench correction curve produced from standards quenched only chemically. The most satisfactory compromise, therefore, is to quench the standards as far as possible in the same way as the samples by including as many as possible of the non-radioactive constituents of the samples, and then to reduce the efficiency of counting further with,

TABLE 7.1. Quenched standards

Each set of figures shows some of the members of a series of quenched standards, prepared by adding to counting vials the materials shown. Scintillation Mixture 2 was used for ¹⁴C standards and Mixture 4 for ³H standards (see text). The two series have been planned on the assumption that the sample of unknown activity was 1 ml aqueous solution. Chloroform has been used as added quenching agent. Instrument set for maximum integral count

Radioactive standard	Vial number	Reagents					dpm	cpm	Counting efficiency
		Scintillation mixture ml	Water ml	Chloroform ml	Toluene ml	Radioactive standard mg			%
¹⁴C–n–hexadecane (2978 dpm/mg)	1*	10	1	0	0·3	9·1	27 100	22 162	81·8
	2	10	1	0·1	0·2	11·1	33 056	25 799	78·0
	3	10	1	0·2	0·1	11·0	32 758	24 496	74·8
	4	10	1	0·3	0	10·1	30 078	21 233	70·6
³H–n–hexadecane (4710 dpm/mg)	1*	10	1	0	0·3	10·3	48 513	14 204	29·3
	2	10	1	0·1	0·2	10·6	49 926	9 690	19·4
	3	10	1	0·2	0·1	10·4	48 984	6 225	12·7
	4	10	1	0·3	0	10·1	47 571	3 969	8·3
Blank	—	10	1	0	0·3	0	—	56	—

* Least quenched standard

for example, chloroform. If the range of quenching is limited, the quench correction curve thus produced will give an adequate correction for the quenching of even those samples which are coloured.

External standard

An external standard is a γ—ray emitter commonly incorporated in the liquid scintillation counter. It is stored in the instrument at a suitable distance from the counting chamber and shielded from it, but when required, it is automatically brought close to the counting vial. Inter-action of γ—rays with an orbital electron of a solvent molecule in the vial may cause ejection of the electron with absorption of the γ—ray photon (*photoelectric effect*) or with lengthening of the γ—ray wave-length (*Compton effect*), or an electron and positron may be generated spontaneously (*pair production*), depending on the energy of the γ—ray; X—rays may also be emitted in the solvent. These interactions give rise to scintillations.

The activity of the external standard is irrelevant for calibration purposes as it is used only to provide a source of radioactivity. Its count rate is first determined in each of a series of quenched standards whose counting efficiency is known, and the count rate of the external standard is then related to this efficiency. The external standard count rate obtained in a sample of unknown activity can then be used to give the efficiency of counting for that sample.

The pulse voltage spectrum produced by an external standard is affected by quenching materials in a similar sort of way to that pro-duced from an internal β—emitting source. The count rate from an external standard may also be particularly affected by the following factors:

1. The position and orientation of the vial in the counting chamber, and the position of the external standard itself. The accuracy of the positioning of the external standard varies in different instruments.
2. The type and dimensions of a vial. For example, we have found a difference of over 3% between count rates of identical samples in vials from different manufacturers.
3. The volume of fluid in a vial. Fig. 7.5 shows the effect of different volumes on both the count rate and the channels ratio (p. 106) of an internal and an external standard; the variation in both values was greater with the external standard than with the internal standard.

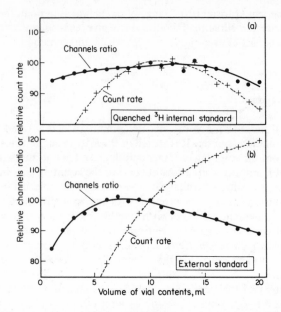

Fig. 7.5. Effect of volume of vial contents on internal and external standard channels ratio compared with effect on corresponding count rate, using scintillation mixture containing chloroform as quenching agent.

(a) Channels ratio and count rate of ³H–n–hexadecane internal standard, 436450 dpm. Relative value of 100 corresponds to a channels ratio of 3·45 and to a count rate of 27000 cpm (counting efficiency, 6%).

(b) Channels ratio and count rate for external standard (¹³³Ba). Relative value of 100 corresponds to a channels ratio of 3·13 and a count rate of 307600 cpm.

Experimental details. Procedure as described in legend to Fig. 4.4. Two analysing channels, A and B, initially set at balance point for internal and external standard respectively with vial content of 10 ml, and discriminators at their widest practicable settings. The upper discriminator of Channel A was lowered to give a count rate equal to about one-third of that in Channel B. Ratio is

$\dfrac{count\ in\ Channel\ B}{count\ in\ Channel\ A}$. *Relative count rate shown is count in Channel B. Scintillation mixture: toluene/chloroform (7:1 v/v) containing 4 g PPO/litre. Counting time, 1 min*

The most commonly used external standards are ¹³³Ba (mainly used in Nuclear-Chicago and Philips instruments), ¹³⁷Cs (mainly used in Beckman and Intertechnique instruments) and ²²⁶Ra (mainly used in Packard and Tracerlab instruments). Some of the differences between the various types of external standard are discussed in Chapter 11.

METHODS OF DETERMINING THE EFFICIENCY OF COUNTING — DIRECT METHOD

There is only one direct method. Although it is the simplest in principle, it is only suitable for small numbers of samples, since with large numbers it is time-consuming and expensive. It is usually referred to as:

Added internal standard or 'spiking' method

This method is based on the assumption that when a radioactive standard is added to a counting vial containing the sample, it is measured with the same efficiency as the sample itself, although under certain conditions this assumption is not always warranted. For adequate statistical accuracy, the total count rate after addition of the standard should be at least ten times that of the sample[106].

Instrument settings

One analysing channel is required.
    Set instrument controls as required for counting the sample.

Procedure

1. Count the vial which contains the sample of unknown activity.
2. Add a precisely measured amount of internal standard and mix thoroughly.
3. Count the vial again.
4. Subtract the first count rate from the second; this gives the count rate attributable to the added standard.
5. Calculate the disintegration rate of the added standard from the known specific activity and the amount added.
6. Calculate the efficiency of counting from the count rate and the disintegration rate of the added standard.
7. Calculate the disintegration rate of the sample from the count rate found in (1) and the efficiency found in (6).

Example

1. Count rate of vial containing sample
      of ^{3}H–labelled compound   10 000 cpm

2. Add 20 mg ^{3}H–n–hexadecane internal
standard, (specific activity 3 500 dpm/mg)
3. Total count rate (sample + standard)   34 000 cpm
4. Count rate of standard   34 000 − 10 000 = 24 000 cpm
5. Disintegration rate of added standard   3 500 x 20 = 70 000 dpm
6. Efficiency of counting $\dfrac{24\,000}{70\,000}$ x 100 = 34·3%
7. Activity of sample   10 000 x $\dfrac{100}{34·3}$ = 29 150 dpm

Advantages of added internal standard method

1. Direct measurement of efficiency of each sample.
2. Simple instrument settings and procedure.
3. Only one analysing channel needed.
4. Valid for colour quenching as well as for chemical and dilution quenching.

Disadvantages of added internal standard method

1. Unsuitable for measuring efficiency for samples on solid supports, or any other samples which are not completely dissolved, since quenching of the sample is then different from that of a standard in solution.
2. Standard must be added separately to each vial, and each vial counted a second time; therefore time-consuming.
3. Count rate of sample cannot be rechecked after addition of standard.
4. Expensive, especially with samples of high activity.
5. Instrument instability may cause error if there is much delay between the two sets of counts, particularly with low counting efficiency (*see* Fig. 4.2).
6. Vial must be opened during procedure, with risk of contamination of equipment.

METHODS OF DETERMINING THE EFFICIENCY OF COUNTING – INDIRECT METHODS

There are three commonly used methods for determining counting efficiency indirectly, and all of these depend upon preliminary measurements made with a set of quenched standards. One method uses an external standard as a source of added radioactivity for determining efficiency. The other two methods relate counting efficiency to the shift of the pulse voltage spectrum that occurs with quenching; in the one case it is the spectrum of the sample in which the shift is measured, and in the other it is the spectrum of the external standard.

External standard counts method

NOTE: When using an external standard it is essential that all vials be identical in composition, dimensions (p. 91) and volume content (*see* Fig. 7.5).

This method is based on the same principle as the added internal standard method. The count rate of the external standard must first be

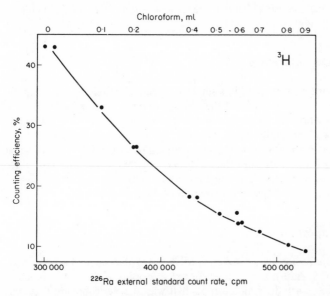

Fig. 7.6. External standard counts method of quench correction using one analysing channel and ^{226}Ra as external standard. Channel was set for maximum differential count of least quenched of a series of quenched ^3H internal standards. Each standard was counted twice, once for 2 min in the absence of the external standard, and once for 1 min in the presence of the external standard. Efficiency of counting for ^3H standard obtained from first count; external standard count rate calculated as the difference between the two count rates. These have been plotted to give a quench correction curve.

Source: ^3H–n–hexadecane, 39 015 dpm/vial. Scintillation mixture: 10 ml toluene/cellosolve (9:1 v/v) containing 4 g PPO/litre, quenched with the amounts of chloroform shown and made up to constant volume with toluene. Instrument, Packard Tricarb Liquid Scintillation Spectrometer Model 3320

determined in each of a series of quenched standards whose efficiency of counting has already been separately determined. The count rate of the external standard is then related to efficiency by means of a quench correction curve such as that shown in Fig. 7.6. If the external standard count rate is now determined in a vial containing a sample of unknown activity, the efficiency can be obtained from the curve. This method of using an external standard, although simple in use, may lead to excessive errors (*see* below).

There are several possible versions of the method, but the applicability of any one depends on the relationship between the γ–ray energy of the external standard and the β–particle energy of the radioisotope being investigated. The manufacturer's instructions should therefore be consulted. For general guidance we give two methods: the first is analogous to the added internal standard method in that it requires only one analysing channel; the second requires two analysing channels but calculations are simpler.

One-channel method

The sample is counted in a single analysing channel, using settings for a differential count, first in the absence of external standard and then in its presence; the difference in count rates gives the count rate due to the external standard.

This method is usually recommended for use with ^{226}Ra. It can also be used with ^{133}Ba, but the relatively short half-life of this radioisotope (7·2 years) causes a fall in count rate of about 1% in 6 weeks, so that the validity of a quench correction curve is short-lived.

Instrument settings

One analysing channel is required. A series of quenched standards is required. Adjust the channel for balance point of the least quenched standard with discriminators at their widest practicable limits (maximum differential count). Maximum integral count settings may be used but the method is then less sensitive.

Procedure

1. Count the least quenched standard in the absence of external standard.
2. Bring the external standard into position and count for 1 min (or other suitable time).

3. Express the counts from (1) and (2) in terms of counts per minute. The difference between them is then the external standard count rate.
4. From the count rate obtained in (1), calculate the counting efficiency and plot against the external standard count rate obtained in (3).
5. Repeat (1) to (4) for each quenched standard to obtain a quench correction curve, as shown in Fig. 7.6.
6. Repeat (1) to (3) for samples of unknown activity and obtain the counting efficiency of each sample from the quench correction curve.
7. Calculate the disintegration rate of each sample from the efficiency obtained in (6) and from its count rate.

Two-channel method

This is only suitable when a large proportion of the pulses derived from the external standard are of higher voltage than those derived from the radioisotope to be measured, as is commonly the case with ^{226}Ra.

One analysing channel is adjusted for counting the radioisotope being investigated, and a second channel for counting pulses derived only from the external standard at voltages above those from the radio-isotope. ^{226}Ra, which gives pulses above the voltage range of ^{3}H and ^{14}C, can be used in this way with both these radioisotopes. ^{133}Ba can only be used in this way with radioisotopes, such as ^{3}H, whose $\beta-$ particles are of low energy, and its use is limited by its relatively short half-life (*see* 'one-channel method' above).

Instrument settings

Two analysing channels are required, one for counting the sample of unknown activity (*sample channel*) and the other for counting the external standard only (*external standard channel*).

A series of quenched standards is required.

Set the controls of the sample channel as required for counting the least quenched standard.

Set the external standard channel for maximum integral count (p. 42). Count the least quenched standard in this channel, and increase attenuation (decrease gain) until the count rate is reduced to about 1% of its original value; it may be necessary finally to raise the setting of the lower discriminator to achieve this. (The value of 1% is arbitrary, and is chosen to ensure that the count rate from the sample in this channel is negligible when compared with that of the external standard.) Bring the external standard into position and count for 1 min; check that the count is large enough (say, at least 100 000 counts) to give the accuracy required over the range of quenching expected.

Procedure

1. Count the least quenched standard for a suitable time in the absence of the external standard. Record the count rate in the sample channel.

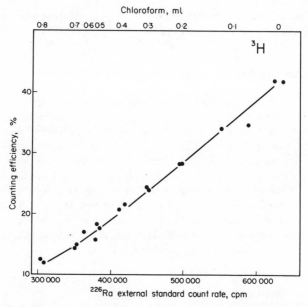

Fig. 7.7. External standard counts method of quench correction using two analysing channels and ^{226}Ra as external standard. One channel (A), set for maximum integral count, was used to count a series of quenched ^3H internal standards, and from these counts their respective counting efficiencies were calculated. The second channel (B) was prepared as described in the text for counting the external standard with minimum interference from the least quenched ^3H standard. Each standard was counted twice, once for 2 min in the absence of the external standard, and once for 1 min in the presence of the external standard. The counting efficiency obtained from Channel A has been plotted against the external standard count rate obtained from Channel B to give a quench correction curve.

Source: ^3H–n–hexadecane, 39870 dpm/vial. Scintillation mixture: 10 ml toluene/cellosolve (9:1 v/v) containing 4 g PPO/litre, quenched with the amounts of chloroform shown and made up to constant volume with toluene. Instrument, Packard Tricarb Liquid Scintillation Spectrometer Model 3320

2. Bring the external standard into position, and count for 1 min (or other suitable time). Record the count rate in the external standard channel.
3. Calculate the efficiency of counting in the sample channel from the count rate obtained in (1) and plot against the external standard count rate from (2).
4. Repeat (1) to (3) for each quenched standard to obtain a quench correction curve, as shown in Fig. 7.7.
5. Repeat (1) and (2) for each sample of unknown activity. From the corresponding external standard count rate for each sample, determine the efficiency of counting from the quench correction curve.
6. Calculate the disintegration rate of each sample from the efficiency obtained in (5) and from its count rate in the sample channel.

Advantages of external standard counts method

1. No addition of material to vial contents required.
2. Sample unaltered, and can therefore be recounted if desired.

Disadvantages of external standard counts method

1. Unsuitable for measuring efficiency for samples on solid supports[17, 28], or any other samples which are not completely dissolved, since in these cases the major cause of quenching is the absorption of β–particle energy by the solid material, and the external standard detects primarily quenching in the solution.
2. Relatively inaccurate[93] owing to possible sources of error peculiar to the use of an external standard (p. 91). Reproducibility of external standard count rate depends on:
   (a) Matching of all vial dimensions.
   (b) Matching of volume of vial contents.
   (c) Reproducibility of relationship between positions of vial and external standard.
3. A quench correction curve based on one type of scintillation mixture does not necessarily apply to another[62].
4. Series of quenched standards required.
5. Inaccuracies may arise when used with plastic vials (p. 156).

Sample channels ratio method[3, 21]

This method relates the shift in the pulse voltage spectrum of the sample to the efficiency of counting. As the spectrum shifts, the count rate over any particular voltage range alters; at higher voltages it falls and at lower voltages it may either rise or fall, depending on the energy of the emitted β–particles (Figs. 7.2 to 7.4). The relationship between the count rate in the high-voltage region of the spectrum and that in the low-voltage region thus changes, and this change can be used as an indication of quenching.

Two analysing channels are used, each of which measures pulses within a particular voltage range; the shift in the pulse voltage spectrum then appears as a change in the ratio (usually known as the *channels ratio*) of the two sets of counts. The efficiency of counting of each of a series of quenched standards is plotted graphically against the channels

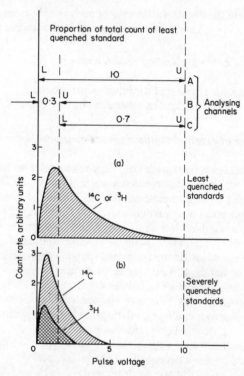

Fig. 7.8. Sample channels ratio method of quench correction; relationship between pulse voltage spectrum and discriminator settings. Diagram hypothetical. (a) Division of the pulse voltage spectrum of the least quenched standard by suitable discriminator settings. Vertical interrupted lines represent these settings (L, lower discriminator; U, upper discriminator). (b) Change in relationship of pulse voltage spectrum to these discriminator settings as a result of quenching. The count rate in the upper voltage range (1·5–10 V) has decreased; in the lower voltage range (0–1·5 V) the count rate of ^{14}C has risen, but that of ^3H has fallen

ratio to give a quench correction curve. If the channels ratio of a sample of unknown activity is found, the counting efficiency may then be obtained from the graph.

There are several convenient ways in which the spectrum may be split between the two analysing channels to give a suitable ratio. Those which have been most commonly described divide the spectrum of the least quenched standard in the proportions 0·3:1·0, 0·3:0·7 (shown diagrammatically in Fig. 7.8) and 0·5:0·5.

We consider the first to be the simplest and in general the most reliable. The others will also be mentioned briefly.

Channels ratio of 0·3:1·0 for least quenched standard

One analysing channel counts all the pulses within the selected voltage range, and a second counts only pulses (0·3 of the total) whose voltages are at the lower end of that range.

Instrument settings

Two analysing channels are required. One, Channel A, is set to count all pulses within the voltage range chosen for counting the sample; the disintegration rate of the sample is therefore calculated from the count rate in this channel. The second, Channel B, is set to count the lower three-tenths of that range.

A series of quenched standards is required.

With the least quenched standard in the counting chamber, adjust both channels either for balance point with the discriminators at their widest practicable limits (maximum differential count), or for maximum integral count (p. 42); if necessary, other settings may be used. Maximum integral count is preferable if the pulse voltages are suitable, since we find that it is more accurate (*see also* Ref. 78) and the settings are simpler.

Make the upper discriminator of Channel B operative if not already so, and reduce its voltage setting until the count rate in that channel is about one-third of the count rate in Channel A (Fig. 7.8). To do this most simply, set the instrument to stop at a predetermined count of, say, 10 000 in Channel A; the upper discrimininator of Channel B should be adjusted until the count in this channel is about 3 000 when the counting stops. The channels ratio for the least quenched standard is then

$$\frac{\text{count in Channel B}}{\text{count in Channel A}} = \frac{3\,000}{10\,000} = 0\cdot3$$

Procedure

1. With the instrument settings obtained as described above, count each of the quenched standards in both channels simultaneously.
2. Determine the efficiency of counting for each standard from the count rate in Channel A and the known disintegration rate of the quantity of radioactive standard present.
3. Calculate the ratio of the count in Channel B to the count in Channel A.
4. Prepare a quench correction curve by plotting the efficiency of counting from (2) against the ratio for each standard from (3). Figs. 7.9 and 7.10 show curves of this type for ^{14}C and ^{3}H respectively, together with the count rate in each channel from which the ratios were derived.

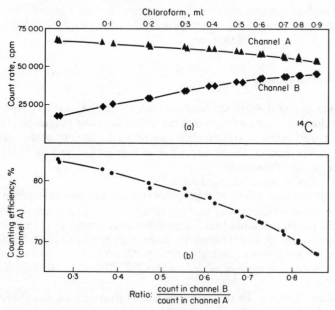

Fig. 7.9. Sample channels ratio method of quench correction; quench correction curve for ^{14}C using ratio of about 0·3:1·0 (see Fig. 7.8) for least quenched standard.

(a) Count rates from a series of quenched standards measured in two channels, A and B, whose discriminators have been set as shown in Fig. 7.8.

(b) Quench correction curve constructed by plotting efficiency of counting in Channel A against the ratio of counts in the two channels.

Source: ^{14}C–n–hexadecane, 79810 dpm/vial. Scintillation mixture: 10 ml toluene/cellosolve (9:1 v/v) containing 4 g PPO/litre, quenched with the amounts of chloroform shown and made up to constant volume with toluene. Counting time, 2 min

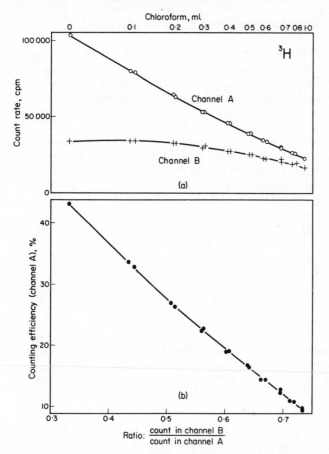

Fig. 7.10. Sample channels ratio method of quench correction; quench correction curve for ^3H using ratio of about 0·3:1·0 (see Fig. 7.8) for least quenched standard. Details as in legend to Fig. 7.9.

Source: ^3H–n–hexadecane, 239 500 dpm/vial. Scintillation mixture: 10 ml toluene/cellosolve (9:1 v/v) containing 4 g PPO/litre, quenched with the amounts of chloroform shown and made up to constant volume with toluene. Counting time, 2 min

5. Count a sample of unknown activity and unknown efficiency in the same way and with the same instrument settings. From the ratio of the counts in the two channels obtain the efficiency of counting from the graph. Calculate the rate of disintegration of the sample from this efficiency and from the count rate in Channel A.

Channels ratio of 0·3:0·7 for least quenched standard

Ideally, three analysing channels are required. Using the least quenched standard two are set as for the previous method, that is, Channel A counts all pulses and the other, Channel B, counts the lower three-tenths of the pulse voltage spectrum. The third channel, Channel C, is set for counting the upper seven-tenths of the pulse voltage spectrum (Fig. 7.8).

The total count rate is obtained from Channel A, and the ratio from the counts in Channels B and C. An example of a graph showing the count rates from the three channels, and a quench correction curve in which the efficiency of counting in Channel A has been plotted against the ratio of counts in the other two channels is shown in Fig. 7.11.

(Alternatively the sum of the counts in Channels B and C may be used in place of the count in Channel A. We do not recommend this, since the voltage values of discriminator settings of different channels seldom match, and the error of the summed counts is therefore greater than that of the single count in Channel A.)

Channels ratio of 0·5:0·5 for least quenched standard

This resembles the last method except that Channels B and C are each set to measure half the counts of the least quenched standard, one channel measuring those from the lower part of the pulse voltage spectrum, and the other those from the upper part.

Choice of ratio

The method using a ratio of 0·3:1·0 is the most satisfactory for general purposes. It has two advantages: (i) one of the two channels measures the total count, and (ii) both the numerator and the denominator of the ratio are maintained at relatively high values with increase in quenching so that error is kept to a minimum.

A ratio derived from a spectrum divided in the proportion of 0·3:0·7 has often been recommended for counting at low efficiencies, particularly with ^3H, in view of the sensitivity to quenching of the count rate in the channel measuring the upper part of the spectrum (*see* counts in Channel C in Fig. 7.11). However, this ratio is not satisfactory if the range of quenching is wide; as can be seen in Fig. 7.11, the count in Channel C falls to low levels when the quenching is much greater than

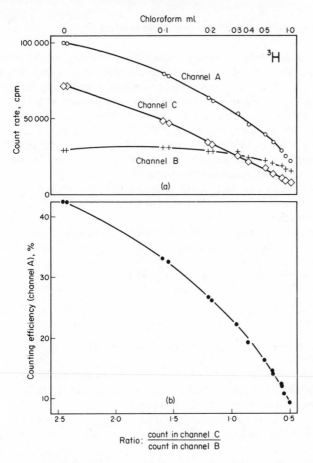

Fig. 7.11. Sample channels ratio method of quench correction; quench correction curve for ³H using ratio of about 0·7:0·3 (see Fig. 7.8) for least quenched standard. Details as in legend to Fig. 7.9. Source and scintillation mixture as in legend to Fig. 7.10

that of the least quenched standard, so that any advantage gained from increased sensitivity may be lost because of increased statistical error.

For similar reasons, the method using a ratio of 0·5:0·5 is recommended only for samples which have little variation in counting efficiency.

The orientation of the ratio in each method is a matter of graphical convenience.

Advantages of the sample channels ratio method

1. This is the only method by which the quenching of radioactive samples on solid supports or in suspension can be compared. This does *not* mean, however, that it is valid for comparing the quenching of material on a solid support or in suspension with the quenching of a standard in solution[17]; the one is usually quenched partly by the support and partly by self-absorption, whereas the other is not. Under these conditions the channels ratio obtained from a standard in solution does not give the true efficiency for counting the undissolved material.
2. Simplicity of counting; the values for the count rate of a sample and for the determination of efficiency are obtained in a single operation.
3. No addition of radioactive material to counting vial is required; sample may thus be recounted after efficiency has been determined.
4. Error is reduced by the use of a ratio.
5. Volume of vial contents is not critical within certain limits (*see* Fig. 7.5).

Disadvantages of the sample channels ratio method

1. Unreliable for samples of low count rate because the accuracy of the ratio depends upon the count rate of the sample itself.
2. Series of quenched standards is required.
3. Unless the range of quenching is narrow, the same quench correction curve cannot be used to correct for both colour and chemical quenching[3, 21, 74].

External standard channels ratio method

NOTE: When using an external standard it is essential that all vials be identical in composition, dimensions (p. 91) and volume content (*see* Fig. 7.5).

This method has been recommended for use with all common types of external standard and is subject to less error than the external standard counts method. The principle is the same as that used in the sample channels ratio method, except that the ratio is obtained from the counts produced by an external standard instead of from the sample.

Ideally, one analysing channel would be used for measuring the radioactive sample, and two more for the measurement of the external standard channels ratio. However, if ^{133}Ba is used as external standard, the lower portion of its energy spectrum may be measured in a channel prepared for counting ^{3}H, and the major part of the spectrum in a channel prepared for ^{14}C. Only two channels are then required for the measurement of ^{3}H, ^{14}C and the external standard, and in addition, it is possible to use the external standard channels ratio method to measure efficiency with samples containing two radioisotopes.

Instrument settings

Two or three analysing channels are required.

If two separate channels are available for counting only the external standard, an external standard channels ratio may be obtained from these channels in an analogous way to that described for the sample channels ratio method (p. 99). Using a blank equivalent to the least quenched standard, the two channels are set at the balance point of the external standard with discriminators at their widest practicable limits (maximum differential count); the upper discriminator of one channel is then lowered until the counts in this channel are about one-third of those in the other to give a suitable ratio.

If the external standard is ^{133}Ba, the necessary ratio can be obtained as just described, but may also conveniently be obtained from two channels which are used primarily for counting ^{14}C and ^{3}H. One of these, which we shall call the *carbon channel*, is used for measuring both ^{14}C and the majority of the pulses derived from the external standard; the other, which we shall call the *tritium channel*, is used for measuring both ^{3}H and the lower-energy portion of the pulse voltage spectrum of the external standard.

A series of quenched standards of the radioisotope to be measured is required.

Adjust the tritium channel for balance point of the least quenched tritium standard with discriminators at their widest practicable limits (maximum differential count). If ^{14}C is the only radioisotope to be counted a ^{3}H standard of similar quenching may be used for preparing this channel.

Adjust the carbon channel for balance point of the least quenched ^{14}C standard with discriminators at their widest practicable limits (maximum differential count). If ^{3}H is the only radioisotope to be counted a ^{14}C standard of similar composition and quenching may be used here; alternatively this channel may be set for maximum differential count of the external standard (using a blank equivalent to the least quenched standard) or for maximum integral count.

These settings give a value of about 3·0 for the ratio

$$\frac{\text{external standard count rate in carbon channel}}{\text{external standard count rate in tritium channel}} .$$

If a suitable ratio (above about 2·0) cannot be obtained using carbon and tritium channels as described, it is advisable to measure the external standard in two separate channels.

Procedure for counting [133]Ba *in carbon and tritium channels*

1. (a) Count the least quenched standard (^{14}C or ^{3}H, whichever is applicable) in the carbon and tritium channels.
   (b) Calculate the efficiency of counting from the count rate in the appropriate channel.
2. (a) Repeat the count with the external standard in position. The count in each channel is now: count of least quenched standard plus count of external standard.
   (b) Subtract the count rate of the least quenched standard in each channel, found in (1a), from the corresponding count rate, found in (2a), to obtain the count rate of the external standard.
3. Calculate the external standard channels ratio, that is,

$$\frac{\text{external standard count rate in carbon channel}}{\text{external standard count rate in tritium channel}} .$$

4. Plot the efficiency obtained in (1b) against the external standard channels ratio calculated in (3).
5. Repeat (1) to (4) for each quenched standard to obtain a quench correction curve as shown in Fig. 7.12.
6. Repeat (1a), (2a), (2b) and (3) for samples of unknown activity. From the external standard channels ratio obtained for each sample determine the efficiency of counting of the sample from the quench correction curve.
7. Calculate the disintegration rate of the sample from the efficiency and the count rate in the carbon channel (for ^{14}C) or in the tritium channel (for ^{3}H).

Advantages of the external standard channels ratio method

1. No addition of material to vial contents required.
2. Sample unaltered and can therefore be recounted if desired.
3. No loss in accuracy with low count rate of sample.
4. Error reduced by using ratio of counts.
5. Less sensitive than external standard counts method to change in volume of vial contents (Fig. 7.5) and in position of vial.

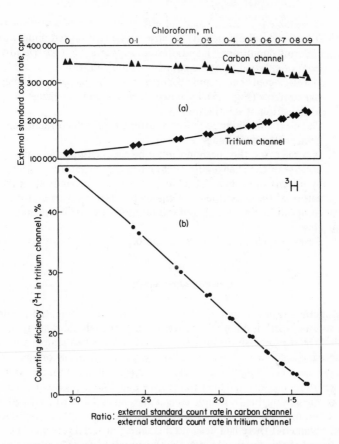

Fig. 7.12. External standard channels ratio method of quench correction. Quench correction curve for quenched ³H internal standards counted first in absence (for 2 min) and then in presence (for 1 min) of external standard (¹³³Ba).

(a) Count rates of external standard measured in a carbon and a tritium channel prepared as described in the text, using a series of quenched internal standards.

(b) Quench correction curve constructed by plotting efficiency of counting ³H standards in tritium channel against the ratio of the external standard counts shown in (a).

Source and scintillation mixture as in legend to Fig. 7.10

Disadvantages of external standard channels ratio method

1. Unsuitable for measurement of efficiency of samples in suspensions (unless the same particle size can be reproduced under all counting conditions[62]), and on solid supports[17, 28] (*see* 'Disadvantages of external standard counts method', above).
2. More affected than sample channels ratio by change in volume of vial contents (Fig. 7.5), by position of vial, and by change in composition of scintillation mixture.
3. Chemical quenching produces a different quench correction curve from colour quenching[74].
4. In some cases, this method cannot be used with plastic vials because of the change of ratio as a result of penetration of wall by solvent and solute[70]. Stability of the ratio depends upon the nature of the constituents of the scintillation mixture and the composition of the vial, and should therefore be investigated before use.
5. Series of quenched standards required.

Extrapolation method

This method is based upon the fact that for a given sample, the smaller the volume which is counted, the less quenching material it contains. Therefore, if progressively smaller volumes of quenched sample are counted, the count rate calculated for each unit volume of original sample rises. This count rate is plotted against volume of sample and the resulting line extrapolated to zero volume, that is, to the point at which quenching due to sample constituents would be zero. This gives the count rate per unit volume of the original sample which would have been obtained if there had been no quenching material present. The efficiency of this count rate may then be obtained from a radioactive internal standard counted in the scintillation mixture used.

This method may be useful when relatively large amounts of a radioactive sample of high specific activity are available, but it is time-consuming when more than a few samples are to be measured. It is only valid if the quenching material is not qualitatively altered by dilution[78].

Instrument settings

One analysing channel is required. Set for maximum integral count only[78]; if the balance point of a quenched sample is used, reduction of quenching by dilution will cause a shift of the pulse voltage spectrum to higher voltages and a consequent fall in count rate per unit volume.

Procedure

1. Add decreasing volumes of the sample of unknown activity to a series of counting vials containing identical volumes of scintillation mixture.

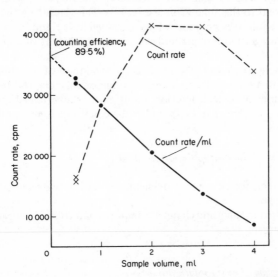

Fig. 7.13. *Extrapolation method of quench correction. A measured quantity of* ^{14}C–n–*hexadecane internal standard (220·4 mg, 2978 dpm/mg) was dissolved in 16 ml chloroform; this represented the sample. Amounts of this sample varying from 0·5 to 4·0 ml were added to 10 ml toluene/cellosolve (1:1 v/v) mixture containing 4 g PPO/litre and made up to 15 ml with toluene. Count rate and count rate per ml of sample have been plotted against sample volume; the plotted line has been extrapolated and cuts the ordinate at a point corresponding to 36350 cpm, which represents the count rate/ml with no interference from sample constituents. A separate quantity of* ^{14}C–n–*hexadecane was dissolved in 10 ml of the above scintillation mixture and made up to 15 ml with toluene. The efficiency of counting this was 89·5% and represents the efficiency with which the sample would be counted with no interference from contained quenching agents. The disintegration rate in 1 ml of sample as given by this method is therefore*

$$36350 \times \frac{100}{89 \cdot 5} = 40615 \; dpm$$

This may be compared with the known disintegration rate calculated from the values given above, i.e., 40267 dpm/ml. Instrument set for maximum integral count

2. Make vials up to constant volume with primary solvent.
3. Determine the count rate of each vial.
4. Plot the count rate per unit volume of original sample against the volume of sample in each vial, as shown in Fig. 7.13.
5. Extrapolate the line joining the plotted values until it cuts the ordinate. This gives a value which corresponds to a count rate which would be given by unit volume of sample in the absence of quenching by sample constituents.
6. Determine the efficiency of counting for an internal radioactive standard in the scintillation mixture alone (made up with primary solvent to the same volume as in (2)). This gives the efficiency of the count rate per unit volume obtained in (5), and from this the disintegration rate per unit volume of sample may be calculated.

As an alternative to (1) to (3), successive aliquots of the sample may be added to a single vial containing scintillation mixture, and the count rate measured after each addition, but it must be ascertained that the overall volume change does not affect the count rate significantly.

Advantages of the extrapolation method

1. Corrects for all types of quenching of radioactive material in solution.
2. Only one analysing channel is required and channel preparation is simple.

Disadvantages of extrapolation method

1. Not suitable for samples in which the radioactive material is not in solution.
2. Time-consuming.
3. Only suitable for samples of high activity.

RELATIVE EFFICIENCY

In certain circumstances it may not be necessary to measure the activity of a sample in terms of disintegration rate. When it is known that a labelled material is unchanged during an investigation, the tracer together with carrier as used in the experiment may take the place of the usual internal standard.

A series of quenched 'standards' using aliquots of tracer plus carrier is prepared in the same way as with internal standard, and from these a quench correction curve of *relative efficiency*, that is, efficiency relative to that of the least quenched 'standard', may be obtained, using any of the indirect methods described. From this curve the count rate of a sample of unknown activity may be compared with the least quenched 'standard'. Since the mass of material in this 'standard' is known, the mass in each sample may be calculated.

CHOICE OF METHOD FOR DETERMINING EFFICIENCY

Only when the material to be counted is in solution is there a choice of method. The most reliable methods are then the internal standard counts and the extrapolation methods, but in practice these are too time-consuming for routine use with large numbers of samples. The most satisfactory method of determining efficiency is then the sample channels ratio method, as this detects most types of quenching. It is, however, only accurate when the count rate of the sample is high. For samples of low count rate, the external standard methods may be satisfactory. The external standard counts method is usually the least accurate[93].

For coloured solutions the added internal standard and extrapolation methods are the most suitable. For both sample and external standard channels ratio methods, colour and chemical quenching agents give different quench correction curves [3, 21, 74]. However, over a limited range of quenching (a relative change in efficiency of about 20%) the curves obtained by the sample channels ratio method coincide, and this method may therefore be used to correct for colour quenching. It may, however, be preferable to decolorise samples first, rather than to rely on a quench correction curve derived from chemical quenching for the measurement of efficiency of counting in the presence of colour quenching. If a quench correction curve must be constructed for coloured samples, it is essential that the material which produces the colour is present in the quenched standards, and that the variation in the quenching of the samples is small, so that little additional chemical quenching of the standards is required.

Determination of the efficiency of counting samples on solid supports or in suspension presents special problems[35]. None of the methods described is satisfactory for quantitative analysis, since the efficiency of counting a standard in solution is no guide to the efficiency of counting a sample which is not in solution. The thickness of the deposited material or the support itself may produce severe quenching quite unrelated to the quenching of a standard in solution[16, 35]. For example, it has been reported that a sample of ^{3}H–labelled material deposited on filter paper was counted with about 6% efficiency, while ^{3}H internal standard dissolved in the scintillation mixture was counted with about 45% efficiency. It has been suggested that the counting efficiency of a sample of unknown activity on a solid support can be determined by the sample channels ratio method using a series of quenched standards prepared from the material under investigation in exactly the same way as the sample. However, we find this unreliable, especially for ^{3}H. A more effective procedure is to elute the radioactive material from the solid support (in the vial) by means of a surface active agent[65], a solubiliser or the appropriate solvent. This will not only raise the counting efficiency but will also make possible its more accurate determination.

NOTES ON THE USE OF THE METHODS

A quench correction curve can only be used to give the efficiency of a sample if that sample is counted with instrument settings identical to those used in the preparation of the curve. These settings must not be transferred to another instrument or even to another channel. After any change involving the instrument, including repair or maintenance, the settings must always be checked and a new curve prepared if necessary.

The determination of efficiency is itself a form of assay and consequently subject to error. The error in an experimentally determined disintegration rate is therefore usually higher than that of the count rate from which it was obtained. For example, in one investigation, we found that the relative standard deviation of the disintegration rate (determined by the sample channels ratio method) was twice that of the count rate of the five identical quenched samples from which it was derived.

For further discussion of the topics mentioned in this chapter, *see* References 28, 78, 83, 93, 94.

CHAPTER EIGHT

Double-isotope analysis

In some experiments certain information can be gained only by following two different groups of molecules or atoms simultaneously through a series of reactions or processes. This can often be achieved by using two different radioisotopes, but it is then necessary to count both of them in the same sample. This is possible in β–particle analysis only if the mean β–particle energies of the two radioisotopes are sufficiently different to enable a proportion of the pulses from one to be counted without interference from those of the other. Whenever possible, however, radioisotopes should be used and measured singly, since the determination of two in the same sample reduces accuracy as a result of cumulative error. In many cases, however, the combined use of two radioisotopes is unavoidable, such as when the ratio of their concentrations is required. Error is particularly large if one of the radioisotopes is measured at low counting efficiency, say 10% or less, and in these circumstances accuracy may be increased by their separation, as for example by a combustion technique (p. 75), before counting.

A method will be described here for the simultaneous counting of ^{14}C and ^3H, since these radioisotopes are the ones most commonly measured together by liquid scintillation counting. However, the principles upon which it is based apply also to other pairs such as ^3H and ^{35}S, ^3H and ^{32}P, and ^{14}C and ^{32}P, and even in certain cases to isotopes of the same element, such as ^{125}I and ^{131}I (*see* Ref. 18).

A large proportion of the pulses derived from ^{14}C are of higher voltage than any of those derived from ^3H, as shown in Fig. 8.1. It is therefore possible for the higher-energy part of the ^{14}C spectrum to be measured in one analysing channel without interference from ^{3}H, and from this measurement the total amount of ^{14}C can be calculated. In a second analysing channel set to measure the lower pulse voltage range, the sum of ^{3}H counts and counts from the lower-energy part of the ^{14}C spectrum are measured. The count rate due to ^{14}C in this channel may be calculated from a knowledge of the total amount of ^{14}C present,

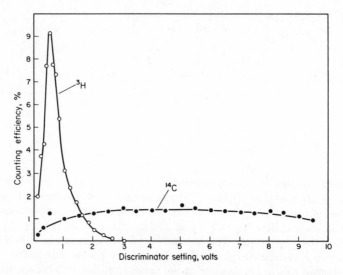

Fig. 8.1. Pulse voltage spectra of ^{14}C and ^{3}H. Points marked represent efficiency of counting plotted against mean discriminator setting, using a window width of 0·1 V, that is, upper discriminator set 0·1 V higher than lower discriminator. Attenuator (gain control) set midway between balance point for ^{3}H and that for ^{14}C.

Sources: ^{14}C–n–hexadecane, 79 810 dpm; ^{3}H–n–hexadecane, 238 400 dpm. Counting time, 1 min

already determined from the measurement in the first channel, and this is then subtracted from the total count rate in the second channel to give the count rate of ^{3}H.

The method will be described in two stages. Initially, the determination of the two radioisotopes will be described on the assumption that quenching is constant for all samples. It will then be shown how this technique can be adapted for the measurement of samples which contain different amounts of quenching material.

Statistical error may be minimised by choosing a ratio of ^{14}C activity to ^{3}H activity which lies between 0·1 and 0·5[51] (*see* Chapter 9).

DETERMINATION OF ACTIVITIES IN SAMPLES WITH EQUAL QUENCHING

Two analysing channels are required; these will be called respectively the *carbon channel* and the *tritium channel* (Fig. 8.2). The carbon channel measures essentially only ^{14}C at pulse voltages above the range covered by the ^3H spectrum. The tritium channel measures mixed ^{14}C and ^3H pulses over a lower pulse voltage range.

Two standards must be prepared, one containing ^{14}C and the other ^3H. To ensure that the amount of quenching material in these standards is the same as that of the samples of unknown activity, they must contain all the constituents of the samples except the radioactive material.

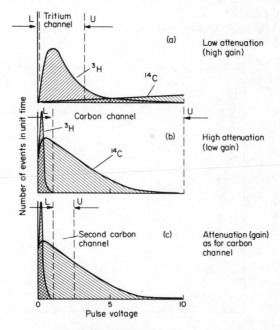

Fig. 8.2. Double-isotope analysis; relationship between pulse voltage spectra and discriminator settings. Vertical interrupted lines represent these settings (L, lower discriminator; U, upper discriminator).

(a) Settings for tritium channel.

(b) Settings for carbon channel.

(c) Settings for second carbon channel in sample channels ratio method of efficiency determination; these differ from those of carbon channel (b) only in the position of the upper discriminator

Preparation of carbon channel (Fig. 8.2(b))

This channel is prepared for the measurement of pulses from the upper
voltage range of the ^{14}C spectrum with minimum interference from ^3H
pulses as follows:

1. Place the ^{14}C standard in the counting chamber.
2. Adjust for balance point with discriminators at their widest
   practicable limits (maximum differential count).
3. Replace the ^{14}C standard by the ^3H standard.
4. Raise the voltage setting of the lower discriminator until the
   count rate derived from the ^3H standard is reduced almost to zero
   in this channel (Fig. 8.3). It may be advisable to retain some of
   the ^3H counts, say 0·1–0·2 per cent, to avoid too great a loss of
   ^{14}C counts[59].
5. Set the lower discriminator of this channel at the value found in
   (4).

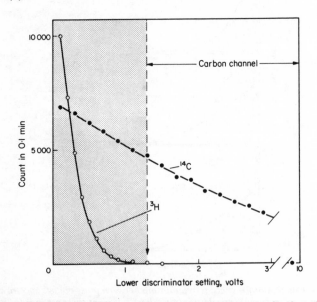

*Fig. 8.3. Double-isotope analysis; setting of lower discriminator of carbon channel.
Channel initially set for maximum differential count of ^{14}C. With ^3H in counting
chamber, lower discriminator was raised until the count rate was almost zero,
thus obtaining required setting of discriminator (shown by arrow). ^{14}C count at
each setting is also shown for comparison.*

    *Sources: least quenched standards used for Figs. 7.9 and 7.10. Counting time,
0·1 min*

6. Replace the ^{3}H standard by the ^{14}C standard.
7. Measure the efficiency of counting for the ^{14}C standard.

Preparation of tritium channel (Fig. 8.2(a))

This channel measures ^{3}H pulses and low-voltage ^{14}C pulses; as many as possible of the interfering ^{14}C pulses must be eliminated without too much loss of ^{3}H pulses as follows:

1. Place the ^{3}H standard in the counting chamber.
2. Adjust for balance point with discriminators at their widest practicable limits (maximum differential count).
3. The upper discriminator must now be adjusted to reduce the contribution of ^{14}C pulses in this channel to an acceptable level. There are various ways of doing this[59]. A simple but satisfactory one is to make the fraction of the initial ^{3}H count which is lost by the adjustment of the discriminator equal to the fraction of the initial ^{14}C count in this channel which is retained (modified from Hendler[49]). This may be carried out as follows (Fig. 8.4), counting in all cases for the same length of time:

    (a) With the ^{3}H standard still in position, reduce the upper discriminator setting in convenient steps, recording the count at each step (Fig. 8.4(b)).
    (b) Repeat with the ^{14}C standard in place of the ^{3}H standard.
    (c) Subtract each ^{3}H count from the initial ^{3}H count and express the difference as a fraction of the initial ^{3}H count (this is the fraction of the initial ^{3}H count which is lost).
    (d) Express each ^{14}C count as a fraction of the initial ^{14}C count in this channel (this is the fraction of the initial ^{14}C count which is retained).
    (e) Plot the values obtained from (c) and (d) against the corresponding voltage settings of the upper discriminator (Fig. 8.4(a)). The curves cross at the discriminator setting required.
    (f) Set the upper discriminator of this channel at the value found in (e).
4. Measure the efficiency of counting for the ^{14}C and ^{3}H standards in this channel.

DETERMINATION OF ACTIVITIES IN SAMPLE OF UNKNOWN ACTIVITY

Using the channel settings already determined, proceed as follows for a sample quenched to the same extent as the standards:

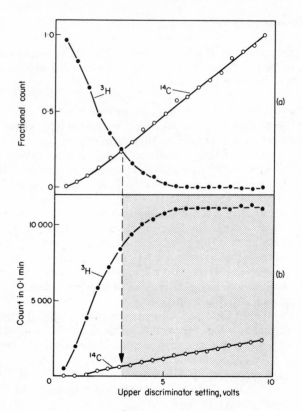

Fig. 8.4. Double-isotope analysis; setting of upper discriminator of tritium channel. Channel initially set for maximum differential count of ^3H. With ^{14}C or ^3H standard in counting chamber, the upper discriminator was reduced in 0·5 V steps, recording the count at each step.

(a) Fractional count at each step, that is,

$$\frac{^{14}C\ count}{Initial\ ^{14}C\ count}\ \ and\ \ \frac{Initial\ ^3H\ count - {}^3H\ count}{Initial\ ^3H\ count}.$$

(b) Count of ^{14}C and of ^3H at each step.

*Crossover point (*vertical interrupted line*) gives upper discriminator setting required.*

Sources: least quenched standards used for Figs. 7.9 and 7.10. Counting time, 0·1 min

1. Determine the count rate of the sample in the carbon and tritium channels.
2. From the count rate in the carbon channel and the counting efficiency for ^{14}C in this channel (step 7, carbon channel above), calculate the disintegration rate of ^{14}C in the sample.
3. From this disintegration rate and the efficiency of counting for ^{14}C in the tritium channel (step 4, tritium channel above) calculate the count rate of ^{14}C in this channel.
4. Subtract from the count rate of the sample in the tritium channel the count rate contributed by ^{14}C in this channel to give the count rate of ^{3}H.
5. From this count rate and the efficiency of counting of ^{3}H (step 4, tritium channel above) calculate the disintegration rate of ^{3}H.

Example

Carbon channel data:

Count rate (^{14}C)	8 000 cpm
Efficiency of counting for ^{14}C	60%
Disintegration rate of ^{14}C	$8\,000 \times \dfrac{100}{60} = 13\,333$ dpm

Tritium channel data:

Count rate (^{14}C + ^{3}H)	19 000 cpm
Efficiency of counting for ^{14}C	10%
Count rate due to ^{14}C	$13\,333 \times \dfrac{10}{100} = 1\,333$ cpm
Count rate due to ^{3}H	$19\,000 - 1\,333 = 17\,667$ cpm
Efficiency of counting of ^{3}H	20%
Disintegration rate of ^{3}H	$17\,667 \times \dfrac{100}{20} = 88\,335$ dpm

DETERMINATION OF ACTIVITIES IN SAMPLES WITH UNEQUAL QUENCHING

The procedure above applies only when the quenching is the same in all samples and standards. In practice, quenching is often varied and unpredictable and its effect on the counting efficiency for each sample must be assessed. Although quenching causes a reduction in count rate, or counting efficiency, of ^{14}C in the carbon channel and of ^{3}H in the tritium channel, it *increases* the count rate, or efficiency, of ^{14}C in the tritium channel, since the shift of the ^{14}C spectrum to lower voltages increases the number of pulses which fall below the upper discriminator level of that channel.

The methods that may be used for determining the three efficiencies for each sample are essentially those that have already been described in Chapter 7. Error, however, is inevitably increased since (1) the efficiency of counting is reduced in narrower windows, (2) instrument instability is magnified by not counting at the balance point of the windows used (discriminator settings are altered *after* the balance point has been determined), (3) ^{14}C data obtained from the carbon channel have to be used in calculations on data from the other channel, and (4) the statistical error is increased when counting two radioisotopes in one channel (p. 134).

Added internal standard method

Instrument settings

A carbon and a tritium channel are prepared as described on p. 117, using ^{14}C and ^{3}H standards whose quenching approximates to that of the samples.

Procedure for each sample

1. Count the sample of unknown activity in both the carbon and tritium channels.
2. Add a known amount of ^{3}H internal standard to the sample.
3. Count the sample again (now with the added ^{3}H standard) in the tritium channel. At the same time check that there is no significant increase in the count rate in the carbon channel.
4. Determine the efficiency of counting ^{3}H in the tritium channel from the additional count rate caused by the added standard and from its known disintegration rate.
5. Add a known amount of ^{14}C internal standard to the sample.
6. Count the sample again (now with both added standards) in both channels.
7. Determine the efficiency of counting ^{14}C in both channels from the additional count rate in each channel caused by the added ^{14}C standard and from its known disintegration rate.
8. Determine the activities of ^{14}C and ^{3}H in the original sample as described on p. 121, using the efficiencies found as described in (4) and (7).

The ^{3}H standard should always be added before the ^{14}C standard[22], since adding ^{14}C first would increase the total count rate in the tritium channel and so reduce the accuracy of measuring added ^{3}H.

The disadvantages of this method are those stated on p. 94; in addition, the vial must now be opened twice.

External standard counts method (Example of method, using ^{226}Ra)

Two sets of quenched standards are required, one of ^{14}C and one of ^{3}H.

Instrument settings

A carbon and a tritium channel are prepared as described on p. 117, using the least quenched ^{14}C and ^{3}H standards. A third channel, the *external standard channel*, is prepared for counting the external standard alone as described on p. 97, with the least quenched ^{14}C standard in the counting chamber.

Procedure

1. Construct three quench correction curves, using the two series of quenched standards separately. That is, determine the efficiency for counting:
   (a) Each ^{14}C standard in the carbon channel.
   (b) Each ^{14}C standard in the tritium channel.
   (c) Each ^{3}H standard in the tritium channel.
   Plot against the corresponding external standard count rate obtained from the external standard channel.
2. For each sample of unknown activity, obtain the external standard count rate from the external standard channel. From the prepared quench correction curves read off against this value the efficiency of counting for:
   (a) ^{14}C in the carbon channel.
   (b) ^{14}C in the tritium channel.
   (c) ^{3}H in the tritium channel.
3. From the count rates in the carbon and tritium channels, determine the disintegration rate of ^{14}C and of ^{3}H in each sample (p. 121), using the efficiencies found as described in (2).

Sample channels ratio method

This method requires a *second carbon channel* (*see* Fig. 8.2) which measures a proportion of the ^{14}C pulses and, in conjunction with the main carbon channel, enables a ratio of ^{14}C counts to be obtained comparable with that described on p. 101. This ratio is then related to the efficiency of counting for both ^{14}C and ^3H.

A series of quenched ^{14}C standards is required, to which ^3H internal standard is added later. (Alternatively, two separate series may be used, one of ^{14}C and one of ^3H. This is inadvisable, since the efficiency of counting the ^3H standards is related to the channels ratio obtained only from the ^{14}C standards, and hence accuracy depends on the ability to reproduce exactly the quenching in both series. Equivalence of quenching in the two sets of standards should be confirmed by ensuring that the external standard channels ratio is the same for each pair.)

Instrument settings

The main carbon channel and the tritium channel are prepared as for samples with constant quenching (p. 117) using the least quenched ^{14}C standard and a separate ^3H standard containing the same non-radioactive constituents.

The second carbon channel is initially prepared in the same way as the main carbon channel. With the least quenched ^{14}C standard in the counting chamber, lower the setting of the upper discriminator of the second carbon channel until the ratio

$$\frac{\text{count rate in second carbon channel}}{\text{count rate in main carbon channel}} \sim 0.3.$$

Procedure

1. Construct two quench correction curves for the efficiency of counting ^{14}C, as shown in Fig. 8.5, using the series of quenched ^{14}C standards. That is, determine the efficiency for counting each ^{14}C standard in
   (a) the main carbon channel,
   (b) the tritium channel
   and plot each value against the corresponding ratio obtained from the counts in the two carbon channels.
2. Obtain a quench correction curve for ^3H, as shown in Fig. 8.6. To do this, add a known amount of ^3H internal standard to each quenched ^{14}C standard, and from the resulting increase in count rate determine the efficiency for counting ^3H in the tritium channel. Plot this efficiency against the ratio of ^{14}C counts obtained at the same time from the carbon channels. (The volume of fluid added is usually too small to affect efficiency significantly.)

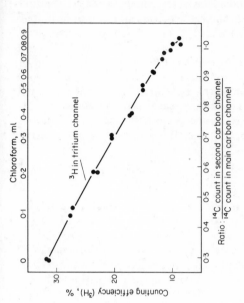

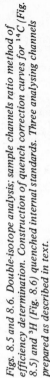

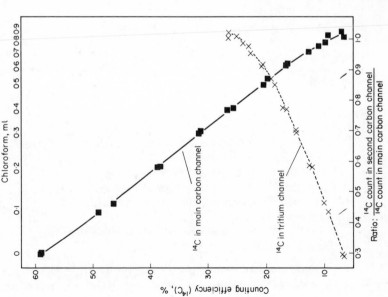

Figs. 8.5 and 8.6. Double-isotope analysis; sample channels ratio method of efficiency determination. Construction of quench correction curves for ^{14}C (Fig. 8.5) and ^3H (Fig. 8.6) quenched internal standards. Three analysing channels prepared as described in text.

Sources: quenched standards used for Figs. 7.9 and 7.10; chloroform as quenching agent. Counting time, 2 min.

Compare quench correction curve in Fig. 8.5 with that in Fig. 7.9; the reduction in counting efficiency in Fig. 8.5 is due to the higher setting of the lower discriminator of the carbon channel. Fig. 8.6 may also be compared with Fig. 7.10. A wide range of quenching has been used for illustration; in practice only a small range of quenching should be used with any one group of instrument settings

3. For each sample of unknown activity obtain the channels ratio
   from the counts in the two carbon channels. From the prepared
   quench correction curves read off against this ratio the efficiency
   of counting for
      (a) ^{14}C in the main carbon channel,
      (b) ^{14}C in the tritium channel,
      (c) ^{3}H in the tritium channel.
4. From the count rates in the main carbon channel and in the
   tritium channel, determine the disintegration rate of ^{14}C and of
   ^{3}H in each sample (p. 121) using the efficiencies found as described
   in (3).

External standard channels ratio method (Example of method, using ^{133}Ba)

Two sets of quenched standards are required, one of ^{14}C and one of ^{3}H.
   A ratio of external standard count rates is derived from the external
standard measured in the carbon and tritium channels. (Alternatively,
the external standard may be counted in the tritium channel and in a
third channel, set either at the balance point of the external standard
or for maximum integral count (p. 42).)

Instrument settings

A carbon and a tritium channel are prepared as for samples with constant quenching (p. 117), using the least quenched ^{14}C and ^{3}H standards.

Procedure

1. Construct three quench correction curves, as shown in Figs. 8.7
   and 8.8, using the two series of quenched standards separately.
   That is, determine the efficiency for counting
      (a) each ^{14}C standard in the carbon channel,
      (b) each ^{14}C standard in the tritium channel,
      (c) each ^{3}H standard in the tritium channel,
   and plot each against the corresponding external standard
   channels ratio, that is,

   $$\frac{\text{external standard count rate in carbon (or third) channel}}{\text{external standard count rate in tritium channel}}.$$

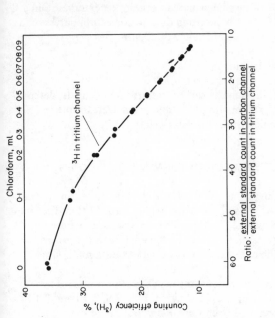

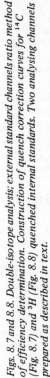

Figs. 8.7 and 8.8. Double-isotope analysis; external standard channels ratio method of efficiency determination. Construction of quench correction curves for ^{14}C (Fig. 8.7) and ^{3}H (Fig. 8.8) quenched internal standards. Two analysing channels prepared as described in text.

Sources: quenched internal standards used for Figs. 7.9 and 7.10; external standard ^{133}Ba; chloroform as quenching agent. Counting time: internal standards, 2 min; external standard, 1 min.

See also comments in legend to Figs. 8.5 and 8.6

2. For each sample of unknown activity obtain the external standard channels ratio. From the prepared quench correction curves read off against this ratio the efficiency of counting for
   (a) ^{14}C in the carbon channel,
   (b) ^{14}C in the tritium channel,
   (c) ^{3}H in the tritium channel.
3. From the count rates in the carbon and tritium channels, determine the disintegration rate of ^{14}C and ^{3}H in each sample (p. 121), using the efficiencies found as described in (2).

CHOICE OF METHOD FOR DETERMINING EFFICIENCY

The criteria for choosing a method of efficiency determination are the same as those described for single-isotope counting (p. 113). If the count rate is high enough the sample channels ratio method is the method of choice, because it detects more types of quenching than the others. For low count rates the external standard channels ratio method may be more suitable.

The measurement of two radioisotopes in the same sample is less accurate than the measurement of one alone. Figs. 7.9, 7.10 and 7.12 and 8.5 to 8.8 were prepared from the same two series of quenched standards; it can be seen that with double-isotope counting the same amount of quenching agent (in this case chloroform) causes a greater reduction in the efficiency of counting for ^{14}C than when it is counted alone. This results in a decrease in the accuracy with which the disintegration rate of ^{14}C, and hence of ^{3}H, is determined. It is therefore important to minimise the range of quenching. If the range of quenching is large, the more severely quenched samples should be counted with settings based on standards quenched to a similar degree; ^{14}C is then counted with higher efficiency in the carbon channel and interferes less in the tritium channel.

For further discussion of the topics mentioned in this chapter, see References 22, 58, 59, 84.

CHAPTER NINE

Random and other errors

The measurement of radioactive material is subject to error from
various sources. Those types of error which are common to all methods
of analysis we shall ignore. However, there are several types of error
which are peculiar to the nature of radioactivity and to its measurement
by liquid scintillation counting, and these we shall describe in this
chapter. We also include a fault-finding chart which, although far from
comprehensive, represents a selection of common problems.

Some errors are predictable and fairly obvious, but others may go
unnoticed and unknown. We ourselves have come across several by
chance during the writing of this book, and it has become clear to us
that nothing must be taken on trust in liquid scintillation counting.
For accurate work meticulous controls and checks, the use of conditions,
reagents and other materials which are identical for each batch of
samples, and the accurate and intelligent use of standards are essential.
The lower the energy of scintillations produced the greater is the likelihood of error. Some errors are unavoidable, but inaccuracies can be
reduced by

(a) using a single isotope rather than a mixture,

(b) using toluene-soluble radioactive materials whenever possible and

(c) in appropriate cases, using ^{14}C rather than ^{3}H.

The type of error which we shall discuss first is the random error
associated with radioactive decay.

RANDOM ERROR OF DECAY[102, 104]

The number of nuclei which will decay in a particular sample of radio-active material in a given time cannot be predicted, since radioactive decay is random. The associated error can, however, be calculated by orthodox statistical methods.

Let us assume that we have a sample containing a radioisotope, such as ^{14}C, whose half-life is so long that any change in the amount of radio-active material present may be ignored. The number of counts (or disintegrations) in each of a series of identical time intervals will then vary in random fashion about a 'true' mean value. This mean value could in theory be found from an infinitely large number of observations. In practice, the probable deviation of a single observation from the mean may be obtained by statistical analysis.

The number of counts (x) recorded in a given time interval is obviously a whole number; repeated determinations made over equal intervals of time will therefore form a distribution varying in discrete steps about a mean. This is a discontinuous distribution, and is best described by a Poisson distribution. When the value of the mean is small, the distribution is asymmetrical, but when the value of the mean becomes greater than about 25 the asymmetry is reduced enough to be ignored, the discontinuity becomes relatively small, and the Poisson distribution approximates to a normal distribution[71].

In a normal distribution, the spread of values on either side of the mean value, $\bar{x}$, shows the variation to which the measurements are subject. This spread may be expressed numerically by the *standard deviation* (s) which, when added to and subtracted from the mean (that is, $\bar{x} \pm s$) gives the limits between which are found 68·3% of the total number of observations (usually referred to as the 68·3% *confidence limits*). Similarly, 95·5% of the values lies within a range contained within two standard deviations each side of the mean ($x \pm 2s$) and 99·7% within three standard deviations ($\bar{x} \pm 3s$). In practice, 95% confidence limits are often used; these are given by $\bar{x} \pm 1·96s$.

For a Poisson distribution, the standard deviation is equal to the square root of the mean value, that is, $s = \sqrt{\bar{x}}$. However, as the number of events in equal intervals of time is made larger, the number of events in any one time interval tends to differ proportionately less and less from the number in any other time interval. If the number of events in any one time interval is large enough, the square root of that number is approximately equal to the square root of the mean number of events, and may therefore be used as a measure of the standard deviation, that is, $s = \sqrt{x}$. This is important in measurements of radioactivity, where often only one measurement is made.

As an example, let us assume that exactly 10 000 counts have been recorded from a radioactive sample. The standard deviation is then $\sqrt{(10\ 000)} = 100$ counts, indicating that the 'true value' has a 68·3% probability of lying between 9 900 counts and 10 100 counts, or that it has a 95·5% probability of lying between 9 800 and 10 200 counts.

The standard deviation may be expressed as a fraction of the count (*relative standard deviation*), $\dfrac{s}{x}$, or as a percentage of the count

(*percentage standard deviation*), $\dfrac{100s}{x}$ or $\dfrac{100\sqrt{x}}{x}$, that is, $\dfrac{100}{\sqrt{x}}$.

This type of expression is often more convenient, since it gives a proportionate error and allows direct comparison of the errors of counts of different magnitude. The percentage standard deviation becomes

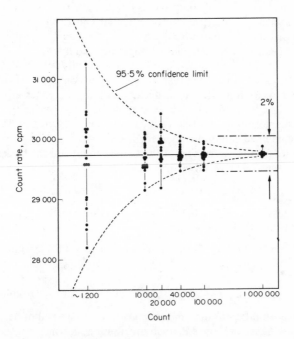

Fig. 9.1. Statistics of radioactive decay. Each vertical series of points represents a count repeated 20 times, except for the count of one million which was repeated only 7 times. Mean count rate, 29 725 cpm. Curved dotted lines represent theoretical 95·5% confidence limits. Arrows show distance on vertical scale corresponding to 2% of mean count rate. Semilogarithmic scale.

Source: sealed ^{14}C–toluene Amersham/Searle reference standard, 31 000 dpm. Instrument set for maximum differential count

smaller as the count increases. Fig. 9.1 shows how this appears in practice. Several series of counts of various magnitudes made on a ^{14}C reference standard have been plotted against count rate. 95·5% confidence limits and the span of an overall 2% of the mean have been included.

It is important to realise that statistically the error of a count depends only on the total number of counts irrespective of the count rate and not on whether they were obtained from one long count or several shorter counts[26]. (The error referred to here applies of course only to that from random causes.)

Background and random error[26, 111]

The normal background count rate is random and the count contributed by the background may therefore be treated statistically in the same way as that due to radioactive decay. When a radioactive sample is measured, the background count is superimposed upon the count derived from the radioactive material whose activity is to be determined. It should go without saying that the greater the count rate of the sample in relation to that of the background, the more reasonable it is to ignore the background. However, when the count rate of the radioactive sample approaches that of the background, a statistical treatment of the combined count rates is needed.

The total count is the sum of the sample count and the background count in the same interval of time, so that

$$\text{sample count} = \text{total count} - \text{background count}$$

Now the variance (s^2) of the difference of two quantities is the sum of the individual variances, so that

$$s^2 \text{ [of sample count]} = s^2 \text{ [of total count]} + s^2 \text{ [of background count]}$$

Therefore

$$s \text{ [of sample count]} = \sqrt{(s^2 \text{ [of total count]} + s^2 \text{ [of background count]})}$$

A number of different determinations of total count and of background count would each follow a Poisson distribution, in which $s = \sqrt{x}$, and therefore $s^2 = x$. It follows that

$$s \text{ [of sample count]} = \sqrt{(\text{total count} + \text{background count})}$$

In other words, the standard deviation of the sample count is equal to the square root of the sum of total count and background count.

This expression applies only to values obtained when the sample and background have been counted for the same length of time. If they have not, as is usually the case, the variances of the counts must be related to a standard time, which we shall take as one minute[26]. If the count in any time interval, t, is taken as x, the counts per minute may be written $x/t \pm s_t/t$, i.e. the variance is now s_t^2/t^2.

We can now express the three variances in terms of one minute, thus:

$$s \text{ [of sample cpm]} = \sqrt{\left(\frac{s^2 \text{ [of total count in time } t_1]}{t_1^2}\right.}$$

$$\left. + \frac{s^2 \text{ [of background count in time } t_2]}{t_2^2}\right)$$

Since $s^2 = x$, this may be written

$$s \text{ [of sample cpm]} = \sqrt{\left(\frac{\text{total count in time } t_1}{t_1^2}\right.}$$

$$\left. + \frac{\text{background count in time } t_2}{t_2^2}\right).$$

As an example, let us assume that a total of 10 000 counts is measured over a period of 100 min (= 100 cpm) and that the background count is 2400 counts measured over 40 min (= 60 cpm).

From these values, sample count rate =

$$100 - 60 = 40 \text{ cpm},$$

and

$$s \text{ [of sample cpm]} = \sqrt{\left(\frac{10\,000}{100^2} + \frac{2400}{40^2}\right)} = \sqrt{(1 + 1 \cdot 5)} = 1 \cdot 6 \text{ cpm}.$$

The count rate of the sample with standard deviation may thus be expressed in terms of one minute, that is, $40 \pm 1 \cdot 6$ cpm.

E^2/B, S^2/B (Figure of merit of instrument)

The expression E^2/B is often used by manufacturers as an indication of the sensitivity of an instrument. It states the relationship between the counting efficiency (E) and the background count rate (B), and indicates the relative contribution of background to the total count under conditions of low count rate. This can be shown in the following way.

The equation which gives the standard deviation of the count of a radioactive sample,

s [of sample count] $= \sqrt{(\text{total count} + \text{background count})}$

may be written

s [of sample count] $= \sqrt{[(\text{sample count} + \text{background count})}$

$+ \text{background count}]$

If sample count and background count in unit time are approximately equal, then

s [of sample count] $\sim \sqrt{(3 \times \text{background count})} \sim \sqrt{3B}$

The relative standard deviation of the sample count is thus:

$$\frac{s \text{ [of sample count]}}{S} = \frac{\sqrt{(3B)}}{S} = \sqrt{\left(\frac{3B}{S^2}\right)}$$

where S is the sample count. This clearly is least when S^2/B is greatest. (Since the count rate of a sample is proportional to the efficiency of counting, the expression E^2/B is then also greatest.) This means that error is reduced by an increase in instrument sensitivity even if accompanied by a proportional increase in background count rate.

Error in double-isotope analysis[22, 51, 59]

Error is considerably greater when measuring a mixture of radioisotopes than when measuring one alone. In the case of a mixture of ^{14}C and ^{3}H, for example, the measurement of ^{3}H is subject to the greater risk of error owing to the encroachment of ^{14}C counts into the tritium channel, either by the relative magnitude of ^{14}C activity or by the shifting of the ^{14}C pulse voltage spectrum to lower voltages by quenching agents. The ^{3}H count is obtained by difference, and, as stated above, the variance of a difference is equal to the *sum* of the individual variances.

Encroachment of ^{14}C counts into the tritium channel can be reduced by

1. Decreasing the proportion of ^{14}C in the sample,
2. Lowering the voltage setting of the upper discriminator of the tritium channel (this will of course reduce the count rate of ^{3}H but there may be an overall improvement in the accuracy of the measurement of ^{3}H),
3. Accurate adjustment of the instrument settings to suit the amount of quenching in each sample.

The error of counting ^{14}C can be reduced by raising its activity, but only at the expense of increased interference with ^{3}H measurement. It has been calculated[51] that the errors of counting ^{14}C and ^3H are roughly equal when the ratio of the activity of ^{14}C to that of ^3H is between 0·1 and 0·5.

BACKGROUND

The background consists of any counts which are not derived from the radioactive material under investigation, and when it is random the statistical treatment described above can be applied. However, there are several other sources of background counts which, because they are non-random, cannot be treated statistically, but can usually be eliminated.

Details of the various types of background counts are as follows:

Random background

This may be derived from the following sources[94]:

1. Cosmic rays. These cause pulses of high voltage, some of which may be excluded by the use of the upper discriminator.
2. Electrical 'noise'. This causes mostly pulses of low voltage, some of which may be excluded by adjustment of the lower discriminator.
3. Local sources. Counts may be contributed by, for example, parts of the instrument and the material of the vials.

If measured over long periods, the count rate from these sources is virtually constant, provided that the instrument is stable and that there are no changes in local radioactive sources.

Erratic background

This is background which varies from one vial to another, and may be derived from:

1. Constituents of the counting vial. The background contributed by a series of vials is only random when all vials are identical. The background count rate from plastic vials is less than that from glass vials, and that from low-potassium vials is less than that from standard glass vials. Where the count rate of the sample approaches that of the background, vials of different types should therefore not be mixed.

2. Contamination of glass vials by radioactivity. This is usually due to inadequate washing and may be avoided by the use of disposable vials.
3. Contamination by radioisotope and scintillator of any part of the instrument which comes into contact with vials.

Transient background

In this type of background the count rate starts at a high level and then gradually falls to a constant value. The pulses are of low voltage. There are two main causes:

1. *Chemiluminescence*[15, 55, 56, 82]

This is a chemically induced fluorescence which occurs particularly at the alkaline pH produced by solubilisers and may persist for many hours. Chemiluminescence induced with a few solubilisers and its progressive reduction with time are shown in Table 6.2. The amount of chemiluminescence depends on the nature of the solubiliser and on the purity of reagents (Table 9.1), and should be tested for with a blank sample containing the materials to be used.

Methods of reducing the count rate due to chemiluminescence are as follows:

1. Use scintillation grade reagents (Table 9.1), and avoid dioxane-based scintillation mixtures.

TABLE 9.1. Effect of grade of toluene on background count rate in the presence of solubiliser ('Soluene', Packard Instrument Ltd.)
In each case 1 ml of solubiliser was added to 10 ml toluene containing 5 g PPO/litre and sample counted for 1 min at times shown. No attempt was made to shield vial from light before initial insertion into instrument, but there was no subsequent exposure. Values shown are approximate. Instrument set for maximum integral count

Grade of toluene*	Count rate (cpm) after				
	1 min	10 min	30 min	60 min	24 hr
Scintillation	1660	50	50	50	50
Rectified	10310	280	100	85	55
A.R.	11450	270	90	70	55
Sulphur-free	60750	1460	400	220	60
A.R. stored unprotected from light	224160	–	–	–	–

* BDH Chemicals Ltd.

2. Delay counting until the count rate has stabilised.
3. Add an acid (such as hydrochloric or acetic) to bring the pH to approximately 7, or add di–*t*–butyl–4–hydroxytoluene (BHT)[56, 57, 81].
4. Raise the setting of the lower discriminator.

Chemiluminescence may be eliminated by strong acidification with hydrochloric acid, but materials such as proteins are then precipitated. However, precipitation can be prevented and a clear homogeneous solution obtained by the addition of a secondary solvent before the acid, the amount required depending on the quantities of other materials present.

2. *Photoluminescence*

This is a fluorescence induced by light. The count rate usually falls to the natural background level fairly quickly, but it has been reported to take as much as an hour[94]. The magnitude of the effect depends on the type of light source which induces it and on the quality of the reagents.

OTHER TYPES OF ERROR

There are many other types of error, most of which cannot be treated statistically. Some of the common ones are listed below.

Errors related to instrument

Instrument drift

The lower the efficiency of counting, the more noticeable will be instrument drift. With a severely quenched sample, it is important to check for drift continuously by means of a radioactive standard quenched to the same efficiency as the sample.

Faulty settings

It is always advisable to check the positions of all controls both by sight and by feel. A control may be at the wrong setting through simple oversight, but all too often a control becomes misaligned as a result of its shifting on its spindle.

Coincidence loss

This is unlikely to be seen unless the count rate is extremely high.

Errors related to vial and contents

Several sources of error associated with vial and contents have already been discussed in Chapter 4. These are:

1. Quenching,

2. Self-quenching,

3. Phase separation (including incomplete mixing of vial contents),

4. Precipitation,

5. Surface adsorption,

6. Change in temperature,

7. Low counting efficiency in relation to instrument drift,

8. The nature of the vial itself (*see also* Chapter 11),

9. Scratches or other marks on vial,

10. Volume of vial contents.

Other sources, already described in this chapter, are contamination of vials and chemiluminescence and photoluminescence.

Variations in the purity of the constituents of different batches of scintillation mixture may affect counting efficiency, background count rate and chemiluminescence. It is therefore advisable to prepare sufficient scintillation mixture in order to provide uniformity in each complete series of samples and standards.

The size and shape of a vial, as well as the material of which it is made, may affect the count rate, particularly of an external standard. Plastic vials should never be used more than once.

The effect of temperature may be important in relation to phase separation and precipitation, particularly when a refrigerated counting well is employed.

FAULT-FINDING

Suggestions are given in Table 9.2 for tracing the causes of some of the common faults which occur in liquid scintillation counting. To use the table two standards are required and each vial must be counted more than once. One of these standards is a vial containing all the constituents of the sample vial but with the radioactive material replaced by a toluene-soluble internal standard and counted under identical conditions; the other is a sealed unquenched reference standard.

The reference standard is used primarily to distinguish between faults in the instrument, which usually require the attention of a service engineer, and faults which are associated with the sample and the operation of the instrument.

For fault-finding chart (Table 9.2), *see* pp. 140–144.

For further discussion of topics mentioned in this chapter, *see* Reference 84.

TABLE 9.2. Fault-finding chart

Suggestions for investigating and treating some faults in liquid scintillation counting. 'Internal standard' represents a vial with contents which are in every respect identical with the vial containing the sample except that the radioactive material is replaced by a toluene-soluble internal standard

Fault in vial containing sample	Observations on		Further observations on vial containing sample	Possible cause of fault	Remedy or prevention
	Sealed reference standard	Internal standard			
Count rate rising (i) slowly	Count rate stable	Count rate may rise	Count rate rises and stabilises after shaking vial	Inadequate mixing of sample and scintillation mixture	Count again after thorough mixing
	Count rate stable	Count rate rising slowly	(a) Fault more marked with low counting efficiency; count rate stabilises in about 30 min	Counting chamber refrigerated; temperature equilibration not reached	Delay counting for at least 30 min
			(b) Toluene-soluble sample in plastic vial	Delay in adding scintillation mixture to toluene-soluble material in plastic vial	Add scintillation mixture before, or immediately after, radioactive material; or use glass vial
	Count rate may rise slightly	Count rate rising slowly	Fault more marked with low counting efficiency; subsequently count rate may fall	Instrument drift	Divide counting time into several short periods; or reduce quenching; or increase activity to reduce counting time
(ii) rapidly	Count rate stable	Count rate may rise	Vial contents milky or separated into two phases	Phase separation of lipid-soluble radioactive material in aqueous sample	Increase proportion of secondary solvent

Fault in vial containing sample	Observations on		Further observations on vial containing sample	Possible cause of fault	Remedy or prevention
	Sealed reference standard	Internal standard			
Count rate falling (i) slowly	Count rate stable	Count rate stable	Count rate rises on shaking vial	Precipitation of radioactive material in vial	Dissolve material; or suspend in gel
	Count rate may fall slightly	Count rate falling slowly	Fault more marked with low counting efficiency; subsequently count rate may rise	Instrument drift	Divide counting time into several short periods; or reduce quenching; or increase activity to reduce counting time
(ii) rapidly	Count rate stable	Count rate may rise	Vial contents milky or separated into two phases	Phase separation of water-soluble radioactive material in aqueous sample	Increase proportion of secondary solvent
	Count rate stable	Count rate initially excessive; falling rapidly	Count rate initially excessive; excess counts from pulses of low voltage	(a) Photoluminescence (count rate stabilises rapidly) (b) Chemiluminescence	Delay counting with samples in dark Delay counting until count rate stable and/or neutralise and/or raise lower discriminator

| Fault in vial containing sample | Observations on | | Further observations on vial containing sample | Possible cause of fault | Remedy or prevention |
	Sealed reference standard	Internal standard			
Count rate lower than expected	Count rate normal	Count rate normal	(a) Count rate discrepancy varies from vial to vial	Adsorption of radioactive material on to vial wall or flat surface (2π counting)	Dissolve or suspend radioactive material
			(b) Solution in vial may not be clear or precipitate may be visible; different efficiencies from sample and external standard channels ratio	Sample not completely dissolved	Dissolve sample adequately; or suspend in gel
	Count rate normal	Count rate may be low	Count rate rises and stabilises after shaking vial	Inadequate mixing of sample and scintillation mixture	Recount after thorough mixing
	Count rate normal	Count rate lower than expected	(a) Plastic vials used	Absorption of light by plastic vial	Determine efficiency; or add secondary scintillator or use glass vials
			(b) Volume of vial contents small; effect more marked with low counting efficiency	Volume of vial contents below optimum	Increase volume of vial contents to satisfactory level with scintillation mixture

Fault in vial containing sample	Observations on		Further observations on vial containing sample	Possible cause of fault	Remedy or prevention
	Sealed reference standard	Internal standard			
	Count rate normal	Count rate low	Channels ratio differs from that of reference standard and may vary from vial to vial	Quenching	Determine efficiency
	Count rate normal	Count rate may rise	Vial contents milky or separated into two phases	Phase separation of water-soluble radio-active material in aqueous sample	Increase proportion of secondary solvent
	Count rate stable; may be low	Count rate stable but low	Count rate stable	Instrument controls wrongly adjusted	Readjust controls
Count rate higher than expected	Count rate normal	Count rate normal or raised	Increase in count rate variable from vial to vial; count rate steady; 'balance sheet' shows excess total activity	Contamination of vial or contents	Detect source of contamination and eliminate; clean vials; prepare fresh sample
	Count rate normal	Count rate initially falling rapidly	Count rate initially falling rapidly; excess counts from pulses of low voltage	(a) Photoluminescence (count rate stabilises rapidly)	Delay counting with samples in dark
				(b) Chemiluminescence	Delay counting until count rate stable and/or neutralise and/or raise lower discriminator

| Fault in vial containing sample | Observations on | | Further observations on vial containing sample | Possible cause of fault | Remedy or prevention |
	Sealed reference standard	Internal standard			
	Count rate of standard and blank raised	Count rate of standard and blank raised	Increase in count rate constant; high background from empty counting chamber	Contamination of counting well with radioactive scintillation mixture	Clean counting chamber
Vial counted for less than designated time	Counted for chosen time	Counted for chosen time	Low count rate	Low count reject setting not switched off	Check low count reject control and adjust as necessary
	Counted for chosen time	Counted for chosen time or vial rejected	Non-standard or faulty vial; swollen plastic vial; vial cap malfitting or not screwed down	Rejection due to malfitting vial	Check vials and replace as necessary
External standard count rate or channels ratio variable	External standard count rate and channels ratio normal	External standard count rate and channels ratio unpredictable	External standard count rate and channels ratio vary from vial to vial	(a) Vials of different shape, size or nature (b) Volume of contents varies from vial to vial	Use uniform vials Make content of all vials up to constant volume
	External standard count rate and channels ratio normal	Progressive change in external standard channels ratio	Progressive change in external standard channels ratio in a given vial	Penetration of wall of plastic vial by solvent and scintillator	Count without delay; or use glass vials

CHAPTER TEN

Danger and protection

There are two main types of hazard associated with the use of radio-isotopes: one is the danger to biological organisms and the other the possible contamination of the environment. Both can be reduced to a minimum by combining common sense with a reasonable understanding of the toxic and physical properties of radioisotopes. Radioactive materials are well known to be potentially dangerous, but the fact that the danger is invisible can lead to negligence in handling, particularly where small amounts are used. Such negligence may be revealed initially only by the distortion of experimental data resulting from contamination of material or equipment.

These hazards should of course be no deterrent against the use of most tracers with appropriate precautions, since these are usually used in very small quantities.

DANGER TO MAN

Irradiation causes ionisation and consequent damage in biological tissues, but it is difficult to obtain adequate data on damage to human tissue from low doses. Much of what is stated about the effects of radiation on man is derived by extrapolation from results of animal experiments, from the therapeutic and diagnostic use of ionising radiation, and from the consequences of serious laboratory accidents and nuclear explosions.

Radioactive material may affect tissues by irradiation from outside the body (*external irradiation*) or from inside (*internal irradiation*).

145

External irradiation

γ-rays are the most harmful type of emission from radioisotopes
situated outside the body, since they readily penetrate all kinds of
matter. α-particles and low-energy β^--particles are less harmful so long
as they remain outside the body, since they penetrate no deeper than
the skin, whose superficial layers are being continually discarded, but
high-energy β^--particles penetrate more deeply. For example, β-particles
emitted by ^{32}P (1·7 max. MeV) have an average range of about 0·8 mm
in the skin, but a few may penetrate 6 or 7 mm. Also the effect of α-
particles is fairly localised because of their short range. In contrast,
γ-rays cause damage which is usually more diffuse because less energy
is transferred per unit distance, but the skin nearest to the source
usually receives the largest effective dose.

Internal irradiation

Radioactive material may enter the body by absorption through the
skin, by ingestion or by inhalation. The effective damage caused by a
given quantity of radioisotope situated in living tissue depends upon
(a) the total energy transferred to the tissue and (b) the volume of
tissue over which the energy is distributed. Taken as a whole, α-particles
are more damaging than β-particles in view of their generally higher
energy and their shorter range; γ-rays are less dangerous since they are
more penetrating and so their energy is diffused more widely, but it
must be remembered that most γ-emitters also emit α- or β-particles.
    The factors which affect the amount of energy transferred to a given
volume of tissue are (a) the quantity of radioisotope, (b) its physical
half-life, (c) its biological half-life (the time taken to eliminate half of a
quantity of material from the body), (d) localisation of the radioactive
material in that particular tissue, and (e) as implied above, the type and
energy of the radiation. An example of damage caused by the localisation of an α-emitter is the bone cancer which developed in painters
of luminous dials as a result of the ingestion of radium.
    The physical half-life and the biological half-life can be combined to
give the biologically more meaningful *effective half-life* (*see* Table 2.3),
which is the time taken for half of the radioisotope initially present to
disappear from the body (by a combination of decay and excretion). The
effective half-life for some radioisotopes, such as ^3H and ^{14}C, cannot be
stated in general terms, since the biological half-life depends on the
molecule into which the radioisotope is incorporated. For example,
$^{14}CO_2$ and $^{14}CO_3{}^{2-}$ are rapidly excreted from the body, whereas ^{14}C–

thymidine incorporated into nucleic acid has a long biological half-life. Because of its short physical half-life and rapid biological turn-over, ^{24}Na, for example, is far less dangerous than ^{90}Sr, which not only has a long effective half-life but is also localised in bone, and is therefore particularly dangerous.

Harmful effects

The harmful effects of radiation on man may be divided into acute and chronic. Acute effects are the result of a large dose of radiation over a short time. Chronic effects are more usually due to repeated exposure to radiation over long periods. In both cases there is no subjective awareness of overdose until the effects of tissue damage become apparent, since the body has no organ by which radiation can be detected.

Acute effects become apparent within a few minutes to several hours after exposure to large doses (say, 3 000–6 000 rads, *see* below) of penetrating radiation (γ–rays or X–rays). These effects do not occur in tracer work in which only microcurie amounts of material are used. The effects consist of general ill-health accompanied by blood disorders such as haemorrhage, gastrointestinal disturbances and other symptoms. High doses of externally administered γ-, X- or β–radiation may produce skin burns. The earliest detectable changes in the body, which are often used routinely to detect acute overdose, are alterations in the blood cells.

Chronic effects are more insidious and in this sense more dangerous than acute effects, because there is no outward biological evidence of overdose until a number of years have passed. The consequences of chronic irradiation, such as cancer, leukaemia, cataract of the eye, or simply shortening of the life span, are irreversible, so far as is known. The action on the gonads is cumulative[36], that is, repeated small doses, however infrequent, are additive, and the total dose is the sum of these. However, it is believed that other parts of the body can recover from small infrequent doses.

DOSE FROM RADIATION

The dose rate of irradiation by γ–rays or X–rays from an unshielded source outside the body may be calculated provided that the activity of the material, the energy of the radiation and the distance from the source are known.

The standard unit of exposure to radiation is the *roentgen* (R), which is a measure of the ionisation of air by X–rays. It is the quantity of γ– or X–rays which, by ionisation, will produce one electrostatic unit of electricity in 1 cc of dry air at standard temperature and pressure.

From the biological point of view, it is more important to relate the exposure to the energy released in tissue, since it is this which determines the amount of tissue damage. Similarly, in chemistry one is interested in the energy deposited in a given mass of material. Furthermore, it is necessary in the study of nuclear radiations to have a unit of energy transfer which can apply not only to γ– and X–rays but to any type of radiation, including protons and neutrons, since all have the same qualitative effect. This unit is the rad (*radiation absorbed dose*) which is related to the absorption of energy by matter and represents the dose of any ionising radiation which results in the liberation of 100 ergs of energy in 1 g of absorbing material. It has been found that 1 roentgen of γ– or X–rays results in the transfer of nearly 100 ergs of energy to 1 g of water or soft biological tissue (this does not apply to hard tissue such as bone). For water or soft tissue, therefore, exposure to 1 roentgen results in a dose of about 1 rad. (The reader may also encounter the rep, which is similar to the rad, but is now obsolete.)

The final concern is with damage to tissue, the amount of which, for a given amount of energy absorbed, is affected by the nature of the radiation. A further unit has therefore been coined, the rem (*roentgen equivalent man*) which is equal to rads multiplied by a damage factor, the RBE (*relative biological effectiveness*) or QF (*quality factor*), thus:

$$\text{rems} = \text{rads} \times \text{RBE (or QF)}.$$

For γ–rays, X–rays and β–particles the RBE (or QF) is 1·0; rems and rads are then identical but doses are usually expressed in rems. With these types of radiation we can say for practical purposes:

Exposure to 1 roentgen produces a dose of 1 rem.

Calculation of exposure rate from a source of γ–rays

The exposure rate at a given distance from a point source of γ–rays with only air between is given[92] as

$$\frac{\text{mCi of activity} \times \text{specific } \gamma\text{–ray constant}}{(\text{distance in cm})^2} \ \text{R/hour.}$$

The specific γ–ray constant (defined as the exposure dose rate produced by the γ–rays from a unit point source of the radionuclide in question at unit distance) may be obtained from the *Radiochemical Manual*[108].

The above units of activity and distance may be replaced by curies and metres respectively, and the value for the constant quoted in the *Manual* must then be divided by a factor of ten.

Maximum permissible dose

'The objectives of radiation protection are to prevent acute radiation effects and to limit the risks of late effects to an acceptable level'[91a]. Maximum doses officially permitted for exposure to radiation (given below and in Table 2.3) are those expected to give a low probability of radiation injury, and their application to any individual must be related to other social and medical factors of that individual[91a]; there is inevitably an element of uncertainty in the values chosen. There is no evidence that there is a minimum threshold dose below which there is no damage to tissues or below which radiation is harmless; even the natural background radiation is believed to contribute towards cancer and genetic mutation[92]. Thus the recommended maximum permissible dose sets a level at which the probability of damage is low enough to be considered acceptable in the light of benefit gained by the employment of the radioactive material.

It seems reasonable to assume that *all* radiation is harmful, and that exposure should be avoided whenever possible by keeping radioactive materials away from the body, by avoiding direct contact with them and by using shielding when necessary. Good laboratory practice should automatically follow and the safety margin will then be large.

External irradiation

The maximum permissible dose of γ-rays or X-rays from a source outside the body varies from organ to organ, since some organs are more sensitive than others. Published maximum permissible doses in the U.K.[25] are as follows:

Blood-forming tissues (bone marrow) and gonads	5 rem/yr
Individual organs	15 rem/yr
Skin and bone	30 rem/yr
Hands, forearms, feet and ankles	75 rem/yr

A relatively higher dose rate is permissible over a shorter period. To put these values in perspective, the mean lethal single whole body dose is about 500 rem and the single dose for sickness symptoms about 150 rem[101], a chest X-ray gives a dose of 0·01–0·1 rem and the dose from natural sources in the U.K. is about 0·1–0·2 rem/yr.

To relate the figures given above to a laboratory example, it can be calculated that exposure to 250 μCi of ^{22}Na at 2 ft for a period of 37 wk or at 1 ft for about 9 wk *continuously* would give a dose of about 5 rem.

Internal irradiation

The maximum amount of a radioactive material which may be present in the body at any one time (*maximum permissible body burden*) is shown in Table 2.3. It should, however, be emphasised that these are *maximum* limits, and should be used as a guide for taking remedial action after accidents and for obtaining some idea of how potentially hazardous work with a particular radioisotope at a particular level may be; in normal practice, any risk of ingestion or absorption of radioactive materials into the body should be avoided.

SAFETY MEASURES

Safety measures are based on the same principles as those used when working with pathogenic bacteria, with the added precautions of shielding and the application of the inverse square law in the presence of γ-emitters. It is important to know where radioactive material is situated, to avoid exposure to it, and to prevent contamination both of the body and of equipment. As Lenihan[61] has succinctly suggested, 'keep away from it . . . ., and interpose some absorbing material'.

The following are some suggestions for keeping irradiation of the individual to a minimum.

1. Avoid direct contact with radioactive materials and keep an adequate distance from unshielded material. Remember the inverse square law, which states that *in vacuo* (and, for practical purposes, in air) the intensity of electromagnetic radiation from a point source decreases as the square of the distance.
2. Use shielding. Normal laboratory glassware stops α- and most β-particles, and is usually adequate for the storage of small amounts of radioisotopes which emit only these particles. Lead shielding is commonly used for γ-emitters.
3. Do not eat, drink or smoke in laboratories used for radioactive work. Dust may contain radioactive material.
4. Do not pipette by mouth.
5. Wash hands after using radioactive material, and wash any radio-active material from the skin immediately, preferably by scrub-

bing with soap and water. This is particularly important for substances readily absorbed through the skin.

6. Wear gloves when handling materials capable of being absorbed through the skin.

7. Avoid breathing radioactive gases.

8. Wear a film badge or other radiation detector on the trunk of the body at all times in a laboratory which contains a γ-emitter or a high-energy β-emitter, even if not working with it. A film badge contains shielded X-ray film from which a rough estimate of dosage may be obtained.

9. Monitor radiation levels as necessary with a Geiger-Müller counter, and check on possible contamination of the hands or benches whenever work with radioactive material is in progress. (^3H is virtually undetectable by this method.)

10. In suspected overdose of some radioactive materials, the amount in the body may be monitored by measuring the activity in urine, by whole-body counting or by localised measurements.

11. In summary, keep exposure to and contact with all radioactive material to a minimum.

General laboratory practice

One of the main precautions to be taken in a laboratory for radioactive work should be the provision of surfaces from which contaminating material may be easily removed. Bench tops may be covered with a disposable covering such as 'Benchkote' (Whatman), which is an absorbent paper backed with polythene, or with brown paper as a cheaper but less effective substitute, but impervious bench surfaces which may be easily cleaned are preferable, since both the cost and the volume of waste material are reduced. Whenever possible, dispensing from stock solutions should be carried out in a drip-tray with an absorbent lining.

Not only the safety of personnel should be considered, but also the contamination of equipment, as this leads to uncontrolled and high background count rates. It is usually advisable to separate working areas, glassware and other equipment into two groups, those used with high levels and those used with low levels of radioactivity; this avoids the risk of transferring radioactive material from high-activity stock solutions to equipment used for tracer work.

Care should be taken also to avoid contamination of equipment by scintillation mixture. Contamination of counting well and conveyor belt with non-radioactive scintillation mixture may increase background by interaction with any contaminating radioactive material. All vials

should therefore be externally free not only from radioactive material
but also from scintillator.

Toxic chemicals[29]

Some of the non-radioactive chemicals used in liquid scintillation
counting are toxic, and they should be used with care. The absorption
of repeated amounts of toluene and xylene by the body depresses the
function of the bone marrow with consequent anaemia, and can cause
fatty degeneration of heart, liver and adrenal glands. Chronic poisoning
by cellosolve causes kidney and liver damage; the fatal acute dose is
reported to be only 10 g[29]. Solubilisers are highly corrosive. Silica-based
gelling agents in powder form probably have the same danger as
powdered silica itself, that is, damage to lung tissue with subsequent
silicosis.

Detection of contaminating radioactivity

A suspected spill of radioactive material should be checked by means of
a Geiger-Müller counter, which can detect γ-rays and medium- or
high-energy β^--particles. It will *not* readily detect β-particles of low
energy, such as those from ^{3}H, since these mostly do not penetrate the
mica window of the instrument, although most instruments will detect
relatively large amounts of ^{14}C.

Low-energy β-emitters on benches and equipment may be detected
by swabbing with cotton-wool which is then placed in scintillation
mixture and counted in the ordinary way in a liquid scintillation
counter.

REGISTRATION AND THE PUBLIC AUTHORITY

Those wishing to use radioactive materials must first be registered with
the appropriate authority, from whom details of regulations governing
the use of such materials may be obtained. In England and Wales this is
the Department of the Environment at the time of writing and in
Scotland it is the Secretary of State. Regulations vary according to
circumstances and it would be out of place to go into detail here; in
larger institutions, however, there is usually a supervising officer from
whom details may be obtained.

The regulations governing waste disposal may vary according to the
region, but, in general, aqueous solutions which contain radioactive

material may, up to a certain limit, often be allowed into the general drainage system. However, organic solvents and solid waste should be collected in separate containers for special disposal.

Publications on safety measures and other aspects of the use of radioactive materials may be obtained from Her Majesty's Stationery Office, London and from the International Atomic Energy Authority, Vienna.

For further discussion of topics mentioned in this chapter, *see* References 2, 25, 48, 60a, 61, 68, 84c, 89, 90, 91, 92 and 104.

CHAPTER ELEVEN

Equipment and materials

THE LIQUID SCINTILLATION COUNTER

The choice of a liquid scintillation counter is often difficult, and it is beyond the scope of this book to compare the merits of every model. There are three main features in which instruments differ: these are (1) type of amplification, (2) the type of external standard and (3) the temperature of the detection unit.

An instrument which has linear amplification is the more complex in use since a different adjustment of the pulse voltage in the analysis circuit is often required for each radioisotope measured. Stability of measurement is, however, said to be greater with a linear amplifier than with a logarithmic one[33], and this is particularly important in the analysis of a mixture of radioisotopes, in which any instrument errors are magnified.

It would be rash to be dogmatic about the choice of external standard. Those usually available are ^{133}Ba (0·4 MeV), ^{137}Cs (0·7 MeV) and ^{226}Ra (up to about 2·0 MeV). ^{226}Ra has the advantage that a large proportion of the pulses generated are at higher voltages than those of ^{3}H and ^{14}C, and it is therefore easily measured independently of these radioisotopes if a separate channel is available. ^{133}Ba, on the other hand, generates pulses whose voltages are of similar order to those of ^{14}C; this makes extra calculation necessary but the sensitivity to quenching agents should be comparable to that of ^{14}C. However, the half-life of ^{133}Ba (7·2 yr) is much shorter than that of either of the other two external standards mentioned, so that its count rate falls by about 1%

in a month; the external standard channels ratio method of quench correction is not affected by this, but the external standard counts method is unsuitable.

In some instruments the counting chamber and associated structures are maintained at a constant low temperature. Before the introduction of coincidence circuits, this used to be particularly important in reducing photomultiplier 'noise'. There is some disagreement over the value of cooling in present-day instruments[86], but it still seems to have three advantages: (1) it provides a constant temperature of photo-multipliers and vial and hence reduces changes in sensitivity due to changes in atmospheric temperature, (2) it reduces the rate of loss of volatile solvents from vials, and (3) it reduces the rate of penetration of certain solvents and solutes into the wall of plastic vials. However, cooling also has disadvantages, chiefly that certain solutes may precipitate at low temperatures.

It is advisable to discuss the following points with as many users of the various makes as possible: (1) facilities provided by an instrument, (2) reliability, (3) delay and costs of servicing and (4) cost and coverage of maintenance contract.

COUNTING VIALS[83]

Vials used for counting are of either glass or plastic and are of a standard size designed to fit commercial automatic liquid scintillation counters. If vials other than those specially made for use in these instruments are used, it is advisable to check that the background count rate is not excessive and that variations in size and in thickness of side or base do not produce variations in the external standard count rate or channels ratio.

Glass vials are of two main types, those made of standard glass and those made of glass with a low ^{40}K content. ^{40}K is a naturally-occurring radioisotope (about 0.1% of all natural potassium) and is the major source of background derived from the glass itself. However, unless the count rate of samples is expected to be in the region of the background count rate it is usually unnecessary to use the relatively expensive low-potassium vials. Glass is the material of choice for vials, although it has certain disadvantages.

Advantages of glass vials

1. Impermeable to solvents.
2. Contents readily visible.
3. Reusable.

Disadvantages of glass vials

1. Scratching of surface may cause variations in efficiency[83].
2. Not usually disposable owing to high cost, so that thorough washing is required after use[44].

The *washing* of glass vials, particularly those which have contained biological material, may present a problem because of the likelihood of adsorption on the vial wall. A suggested procedure is as follows. Empty the vial and rinse thoroughly with water (if the radioactive material was water-soluble), then with acetone to remove most of the non-polar solvents and solutes. Soak in a surface active agent, such as Decon 90 (A. Gallenkamp Ltd.) or RBS 25 (Chemical Concentrates Ltd.), following the manufacturer's instructions. Rinse several times in warm water and then in distilled water before drying. Alternative methods have been described[44, 83]. (A dishwasher with suitable racks is probably an improvement on washing by hand.) The effectiveness of washing should be checked by filling randomly chosen vials with scintillation mixture and measuring the background count rate.

Plastic vials eliminate the need for washing, since they must be discarded after use, but their disadvantages may outweigh this. Their properties depend upon the material of which they are made. They should therefore be tested for errors due to permeability under the conditions in which they are to be used.

Advantages of plastic vials

1. Cheaper to buy than glass vials.
2. Washing up avoided, since a new vial must be used for each count.
3. Give lower background count rate than glass vials.

Disadvantages of plastic vials

1. Toluene and toluene-soluble substances may migrate into the vial wall. When this happens it has the following consequences:
   (a) There is a slow loss of content and fall in count rate as shown in Fig. 11.1.
   (b) The instrument may be contaminated by migration of materials through the vial wall. (For obvious reasons we have not confirmed this, although we have found evidence for permeation of hexadecane through the wall of a polyethylene vial.)

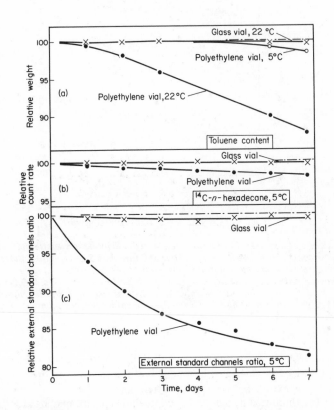

Fig. 11.1. Effect of permeability of polyethylene vials to scintillation mixture constituents.

(a) Changes in weight of counting vial initially containing 10 ml toluene.

(b) Changes in relative count rate of ^{14}C–n–hexadecane in 10 ml toluene containing 4 g PPO/litre. Count rate about 30 000 cpm. Counting time, 1 min. Similar changes can be shown with ^{3}H–n–hexadecane as source.

(c) Changes in external standard channels ratio using [133]Ba, determined as described in Chapter 7 with vials containing 10 ml toluene and 4 g PPO/litre. (In contrast, sample channels ratio for ^{14}C–n–hexadecane in polyethylene vials changed not more than 3% in 7 days.) Sources of vials with standard padded screw caps, Packard Instrument Ltd.

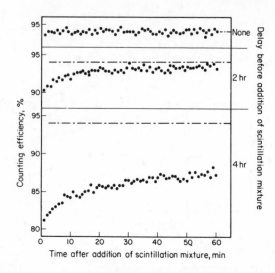

Fig. 11.2. Effect of delay in the addition of scintillation mixture to toluene-soluble internal standard (^{14}C–n–hexadecane) in polyethylene vial. Vial maintained at about 22°C for time stated on right. Scintillation mixture (10 ml toluene containing 4 g PPO/litre) at about 5°C then added and counting started immediately in counting chamber at this temperature. Differences in counting efficiency at time zero due to absorption of standard into vial wall; subsequent changes due to diffusion into the solvent. Activity of standards between 90 000 and 110 000 dpm. Counting time, 1 min. Instrument set for maximum integral count

    (c) There may be softening and swelling of the plastic which, as we have found with some polyethylene vials, can prevent the lowering of a vial into the counting chamber.

    (d) Undiluted hexadecane (or toluene) standard may be absorbed into the vial wall and not all be released into the scintillation mixture, as shown in Fig. 11.2 for polyethylene vials. It follows that when radioactive standard is dispensed into such a vial, scintillation mixture should be added immediately. This precaution should also be taken with any other material that may be absorbed by polyethylene.

    (e) There may be a progressive change in external standard count rate[70] and channels ratio[54] (Fig. 11.1); the rate of change depends upon the solvent and the scintillator. A change in the sample channels ratio has also been reported[70].

2. Contents are not easily visible, so that non-uniformity of mixture may go unnoticed.

3. Efficiency of counting may be a few per cent lower than with glass vials owing to absorption of light; this is worse the lower the efficiency and may be improved by the presence of secondary scintillator.
4. The overall cost of using plastic vials, because they must be discarded after use, is much higher than that of using glass vials.

Small glass specimen tubes provide an economical alternative to standard vials if used in conjunction with a gelling agent and aqueous sample as described on p. 74.

Vial caps which are made of plastic may be washed and with caution used again. Caps containing foil-coated inserts should not be used more than once unless the inserts are renewed.

Carrying trays for vials

For carrying large numbers of vials, metal carrying racks, such as Freezfile Units (Luckham Ltd.) are particularly useful. They are more compact than the cardboard boxes in which vials are usually supplied and can be washed after use to remove contamination.

PIPETTES

Radioactive solutions and scintillation mixture reagents should not be pipetted by mouth. Manually operated or semi-automatic pipettes or dispensers should be used instead, in spite of their extra cost. The following are a few of the types available.

Pipette filler

This is a rubber bulb which fits over the end of a standard glass pipette. It is cheap and simple but relatively slow in use. Its chief disadvantage is that fluid is easily sucked into the bulb, which is difficult to wash out.

Glass syringe

A simple glass syringe is cheap and simple for dispensing solvents, if somewhat slow in use.

Semi-automatic dispenser

This allows aliquots of fluid to be transferred easily and rapidly from a bulk source. There are two main types.

1. An automatic bulb dispenser which consists of a reservoir bottle surmounted by a smaller reservoir of fixed volume from which the fluid is poured.
2. An automatic syringe dispenser which consists of a reservoir bottle surmounted by a syringe and plunger system which can be adjusted to vary the amount dispensed. Rinsing with pure solvent after use may be advisable to prevent jamming. The reproducibility of volumes dispensed by this type is usually considerably better than with type (1) above.

Semi-automatic pipette with manual plunger operation

This type of pipette is convenient and rapid in use. The solution is sucked into a disposable tip so that the pipette itself does not come into contact with radioactive material. A small but progressive error appears to develop as the pipette is warmed in the hand. Some types have an internal rubber sealing washer which may be damaged by toluene vapour. An example of a semi-automatic pipette is the Eppendorf microlitre pipette (Anderman & Co.) which is provided with a statistical analysis of accuracy which we have found to be reliable.

It is important to be certain before purchase that the materials of which any dispenser is made, particularly washers, valves, etc., are resistant to the solvents to be used. An all-glass type is preferable.

An alternative to the use of micropipettes for dispensing small volumes of radioactive standards is to measure them by weight. A short length of fine-bore polythene tubing is fitted over the needle of a 1 ml syringe; liquid is drawn into this without contamination of the needle or syringe and dispensed dropwise.

CHEMICALS

As mentioned earlier in this book, high-grade chemicals are not always necessary for liquid scintillation counting. However, before buying large quantities of a cheap solvent, it should be tested for (1) background count rate, (2) quenching effect and (3) chemiluminescence, especially if solubilisers are to be used.

It is advisable to keep all reagents in dark bottles to reduce the risk of degradation to materials which cause quenching or chemiluminescence.

For further discussion on the choice of liquid scintillation counting equipment and chemicals, *see* References 84a and 84b.

References

Numbers marked with an asterisk indicate a reference of a general nature.

1. ACKERMAN, M. E., DAUB, G. H., HAYES, F. N. and MACKAY, H. A., 'The Photo-oxidation of 2,5–Diphenyloxazole (PPO)', *Organic Scintillators and Liquid Scintillation Counting* (D. L. Horrocks and C.-T. Peng, Eds), 315–325, Academic Press, New York (1971)

2*. ALEXANDER, P., *Atomic Radiation and Life*, Penguin Books, Harmondsworth (1957)

2a. ASHCROFT, J., 'Gamma Counting of Iodine^{125} using a Metal-loaded Liquid Scintillator', *Anal. Biochem.*, 37, 268–275 (1970)

3. BAILLIE, L. A., 'Determination of Liquid Scintillation Counting Efficiency by Pulse Height Shift', *Int. J. appl. Radiat. Isotopes*, 8, 1–7 (1960)

4. BASCH, R. S., 'An Improved Method for Counting Tritium and Carbon–14 in Acrylamide Gels', *Anal. Biochem.*, 26, 184–188 (1968)

5. BAYLY, R. J. and CATCH, J. R., *The Stability of Labelled Organic Compounds*, Review 3, The Radiochemical Centre, Amersham, England (1968)

6. BAYLY, R. J. and EVANS, E. A., 'Stability and Storage of Compounds Labeled with Radioisotopes', *J. labelled Compounds*, 2, 1–34 (1966)

7. BAYLY, R. J. and EVANS, E. A., 'Storage and Stability of Compounds Labeled with Radioisotopes. II', *J. labelled Compounds*, 3 (Supplement 1), 349–379 (1967)

8. BAYLY, R. J. and EVANS, E. A., *Storage and Stability of Compounds Labelled with Radioisotopes*, Review 7, The Radiochemical Centre, Amersham, England (1968)

9. BELL, R. P., *The Proton in Chemistry*, Methuen, London (1959)

10. BENEVENGA, N. J., ROGERS, Q. R. and HARPER, A. E., 'Variations in the Recovery of Carbon—14 in Colored Samples Treated with Peroxide', *Anal. Biochem.*, **24**, 393—396 (1968)

11*. BIRKS, J. B., *An Introduction to Liquid Scintillation Counting*, Koch-Light, Colnbrook, England (undated)

12*. BIRKS, J. B., *The Theory and Practice of Scintillation Counting*, International Series of Monographs on Electronics and Instrumentation, 27, Pergamon Press, Oxford (1964)

13. BOSSHART, R. E. and YOUNG, R. K., 'Preparation of Gas Samples for Liquid Scintillation Counting of Carbon—14', *Anal. Chem.*, **44**, 1117—1121 (1972)

14*. BRANSOME, E. D. (Ed), 342—346, Grune and Stratton, New York (1970) *Counting*, Proceedings of the International Symposium on the Current Status of Liquid Scintillation Counting, held at the Massachusetts Institute of Technology, March 31—April 3, 1969, Grune and Stratton, New York (1970)

15. BRANSOME, E. D. and GROWER, M. F., 'Detection and Correction of Chemiluminescence in Liquid Scintillation Counting', *The Current Status of Liquid Scintillation Counting* (E. D. Bransome, Ed), 342—346, Grune and Stratton, New York (1970)

16. BRANSOME, E. D. and GROWER, M. F., 'Liquid Scintillation Counting of (^{3}H) and (^{14}C) on Solid Supports: a Warning', *Anal. Biochem.*, **38**, 401—408 (1970)

17. BRANSOME, E. D. and GROWER, M. F., 'Local Absorption of Low Energy Betas by Solid Supports: a Problem in Heterogeneous Counting' *Organic Scintillators and Liquid Scintillation Counting* (D. L. Horrocks and C.-T. Peng, Eds), 683—686, Academic Press, New York (1971)

18. BRANSOME, E. D. and SHARPE, S. E., 'Measurement of ^{131}I and ^{125}I by Liquid Scintillation Counting', *Anal. Biochem.*, **49**, 343—352 (1972)

19. BRAY, G. A., 'A Simple Efficient Liquid Scintillator for Counting Aqueous Solutions in a Liquid Scintillation Counter', *Anal. Biochem.*, **1**, 279—285 (1960)

20. BRAY, G. A., 'Determination of Radioactivity in Aqueous Samples', *The Current Status of Liquid Scintillation Counting* (E. D. Bransome, Ed), 170—180, Grune and Stratton, New York (1970)

21. BUSH, E. T., 'General Applicability of the Channels Ratio Method of Measuring Liquid Scintillation Counting Efficiencies', *Anal. Chem.*, **35**, 1024—1029 (1963)

22. BUSH, E. T., 'Liquid Scintillation Counting of Doubly-labeled Samples. Choice of Counting Conditions for Best Precision in Two-channel Counting', *Anal. Chem.*, **36**, 1082–1089 (1964)

23. CATCH, J. R., *Purity and Analysis of Labelled Compounds*, Review 8, The Radiochemical Centre, Amersham, England (1968)

24. CHAPMAN, D. A., 'Determination of Carbon–14, Hydrogen–3, and Sulfur–35 in Rubber Vulcanizates. A New Degradation Method for Liquid Scintillation Counting', *Anal. Chem.*, **43**, 1242–1245 (1971)

25. *Code of Practice for the Protection of Persons Exposed to Ionising Radiations in Research and Teaching*, H.M.S.O., London (1968)

25a. COLLINS, C. J. and BOWMAN, N. S., *Isotope Effects in Chemical Reactions*, Van Nostrand Reinhold, New York (1970)

26. COLQUHOUN, D., *Lectures on Biostatistics*, Clarendon Press, Oxford (1971)

27. COSOLITO, F. J., COHEN, N. and PETROW, H. G., 'Simultaneous Determination of Iron–55 and Stable Iron by Liquid Scintillation Counting', *Anal. Chem.*, **40**, 213–215 (1968)

28. DAVIES, J. W. and HALL, T. C., 'Liquid Scintillation Counting Methods for Accurate Assay of Beta Radioactivity in Biological Experiments', *Anal. Biochem.*, **27**, 77–90 (1969)

28a. DEME, S., *Semiconductor Detectors for Nuclear Radiation Measurement*, Hilger, London (1971)

29. DREISBACH, R. H., *Handbook of Poisoning: Diagnosis and Treatment*, 7th edn, Lange Medical Publications, Los Altos (1971)

30. DUNN, A., 'Interference of Tissue Solubilizers with Liquid Scintillation Counting Fluors', *Int. J. appl. Radiat. Isotopes*, **22**, 212 (1971)

31. ELTON, L. R. B., *Introductory Nuclear Theory*, 2nd edn, Pitman, London (1965)

32*. FEINENDEGEN, L. E., *Tritium-labeled Molecules in Biology and Medicine*, Academic Press, New York (1967)

33. FRANK, R. B. (Nuclear-Chicago Ltd.), pers. comm.

34. FUCHS, A. and DE VRIES, F. W., 'Comparison of Methods for the Preparation of ^{14}C–labelled Plant Tissues for Liquid Scintillation Counting', *Int. J. appl. Radiat. Isotopes*, **23**, 361–369 (1972)

35. FURLONG, N. B., 'Liquid Scintillation Counting of Samples on Solid Supports', *The Current Status of Liquid Scintillation Counting* (E. D. Bransome, Ed), 201–206, Grune and Stratton, New York (1970)

36*. GLASSTONE, S., *Sourcebook on Atomic Energy*, 3rd edn, Van Nostrand, Princeton (1967).

37. GOODMAN, D. and MATZURA, H., 'An Improved Method of Counting Radioactive Acrylamide Gels', *Anal. Biochem.*, **42**, 481–486 (1971)

38. GORSUCH, T. T., *Radioactive Isotope Dilution Analysis,* Review 2,
    The Radiochemical Centre, Amersham, England (1968)
39. GREENE, R. C., 'Heterogeneous Systems: Suspensions', *The Current
    Status of Liquid Scintillation Counting* (E. D. Bransome, Ed),
    189–200, Grune and Stratton, New York (1970)
40. GRIFFITHS, M. H. and MALLINSON, A., 'A Furnace for Combustion
    of Biological Material containing Tritium and Carbon–14
    Labeled Compounds', *Anal. Biochem.*, 22, 465–473 (1968)
41. GROWER, M. F. and BRANSOME, E. D., 'Liquid Scintillation Count-
    ing of Macromolecules in Acrylamide Gel', *The Current Status of
    Liquid Scintillation Counting* (E. D. Bransome, Ed), 263–269,
    Grune and Stratton, New York (1970)
42*. *Guide for Users of Labelled Compounds*, The Radiochemical
    Centre, Amersham, England (1972)
43. GUPTA, G. N., 'A Simple In-vial Combustion Method for Assay of
    Hydrogen–3, Carbon–14, and Sulfur–35 in Biological, Bio-
    chemical, and Organic Materials', *Anal. Chem.*, 38, 1356–1359
    (1966)
44. HALL, T. C. and COCKING, E. C., 'High-efficiency Liquid-scintillation
    Counting of ^{14}C–Labelled Material in Aqueous Solution and
    Determination of Specific Activity of Labelled Proteins', *Bio-
    chem. J.*, 96, 626–633 (1965)
45. HANSEN, D. L. and BUSH, E. T., 'Improved Solubilization Procedures
    for Liquid Scintillation Counting of Biological Materials', *Anal.
    Biochem.*, 18, 320–332 (1967)
46. HARDCASTLE, J. E., HANNAPEL, R. J. and FULLER, W. H., 'A Liquid-
    scintillation Technique for the Radioassay of Calcium–45', *Int.
    J. appl. Radiat. Isotopes*, 18, 193–199 (1967)
47. HAVILAND, R. T. and BIEBER, L. L., 'Scintillation Counting of ^{32}P
    without Added Scintillator in Aqueous Solutions and Organic
    Solvents and on Dry Chromatographic Media', *Anal. Biochem.*,
    33, 323–334 (1970)
48*. *The Hazards to Man of Nuclear and Allied Radiations*, H.M.S.O.,
    London (1960)
49. HENDLER, R. W., 'Procedure for Simultaneous Assay of Two β–
    Emitting Isotopes with the Liquid Scintillation Counting
    Technique', *Anal. Biochem.*, 7, 110–120 (1964)
50. HERBERG, R. J., 'Determination of Carbon–14 and Tritium in Blood
    and Other Whole Tissues. Liquid Scintillation Counting of
    Tissues', *Anal. Chem.*, 32, 42–46 (1960)
51. HERBERG, R. J., 'Statistical Aspects of Double Isotope Liquid
    Scintillation Counting by Internal Standard Technique', *Anal.
    Chem.*, 36, 1079–1082 (1964)
52. HORROCKS, D. L. and PENG, C.-T., (Eds), *Organic Scintillators and*

Liquid Scintillation Counting, Proceedings of the International
Conference on Organic Scintillators and Liquid Scintillation
Counting, held at the University of California, San Francisco,
July 7–10, 1970, Academic Press, New York (1971)

53. IUPAC, 1957 Report of the Commission on the Nomenclature of
Inorganic Chemistry, 'Nomenclature of Inorganic Chemistry', *J.
Am. Chem. Soc.*, **82**, 5523–5544 (1960)

54. JOHANSON, K. J. and LUNDQVIST, H., 'A Pitfall in the Use of Polyethylene Vials in Liquid Scintillation Counting', *Anal. Biochem.*,
50, 47–49 (1972)

55. KALBHEN, D. A., 'Chemiluminescence as a Problem in Liquid
Scintillation Counting', *The Current Status of Liquid Scintillation
Counting* (E. D. Bransome, Ed), 337–341, Grune and Stratton,
New York (1970)

56. KALBHEN, D. A. and REZVANI, A.,'Comparative Studies on Sample
Preparation Methods, Solutes and Solvents for Liquid Scintillation
Counting', *Organic Scintillations and Liquid Scintillation Counting*
(D. L. Horrocks and C.-T. Peng, Eds), 149–167, Academic Press,
New York (1971)

57. KEARNS, D. S., 'Reduction of Chemiluminescence in Both Refrigerated and Ambient Temperature Liquid Scintillation Counting',
Int. J. appl. Radiat. Isotopes, **23**, 73–77 (1972)

58*. KOBAYASHI, Y. and MAUDSLEY, D. V., 'Practical Aspects of Liquid
Scintillation Counting', *Methods of Biochemical Analysis* (D.
Glick, Ed), **17**, 55–133, Interscience Publishers, New York
(1969)

59. KOBAYASHI, Y. and MAUDSLEY, D. V., 'Practical Aspects of Double
Isotope Counting', *The Current Status of Liquid Scintillation
Counting* (E. D. Bransome, Ed), 76–85, Grune and Stratton,
New York (1970)

60. LARSSON, S. 'Low Level Tritium and Carbon–14 Determination',
Anal. Biochem., **50**, 245–254 (1972)

60a*. LAWRENCE, C. W., *Cellular Radiobiology* (Studies in Biology No. 30),
Arnold, London (1971)

61*. LENIHAN, J. M. A., *Atomic Energy and its Applications*, 2nd edn,
Pitman, London (1966)

62. *Liquid Scintillation Workshop*, Notes 1–4 (Tracerlab G. B. Ltd.),
ICN Pharmaceuticals Ltd., Hersham, England (1971)

63. McEVOY, A. F., DYSON, S. R. and HARRIS, W. G., 'Compatibility of
Commercial Tissue Solubilizers with Three Primary Scintillation
Fluors', *Int. J. appl. Radiat. Isotopes*, **23**, 338 (1972)

64. McEVOY, A. F. and HARRIS, W. G., 'Evaluation of Liquid Scintillation
Systems for the Assay of Tritiated Inulin', *Anal. Biochem.*, **43**,
123–128 (1971)

65. McKENZIE, R. M. and GHOLSON, R. K., 'Liquid Scintillation Counting of ^{14}C– and ^{3}H–Labeled Samples on Solid Supports: a General Solution of the Problem', *Anal. Biochem.*, **54**, 17–31 (1973)

66. MAHIN, D. T. and LOFBERG, R. T., 'A Simplified Method of Sample Preparation for Determination of Tritium, Carbon–14, or Sulfur–35 in Blood or Tissue by Liquid Scintillation Counting', *Anal. Biochem.*, **16**, 500–509 (1966)

67. MANTEL, J., 'The Beta Ray Spectrum and the Average Beta Energy of Several Isotopes of Interest in Medicine and Biology', *Int. J. appl. Radiat. Isotopes*, **23**, 407–413 (1972)

68*. MARTIN, A. and HARBISON, S. A., *An Introduction to Radiation Protection*, Chapman and Hall, London (1972)

69. MAURER, H. R., *Disc Electrophoresis and Related Techniques of Polyacrylamide Gel Electrophoresis*, Walter de Gruyter, New York (1971)

70. MUELLER, E. B., 'An Investigation of Homogeneous Counting Systems for Aqueous Inorganic Salts and Acids', *The Current Status of Liquid Scintillation Counting*, (E. D. Bransome, Ed), 181–188, Grune and Stratton, New York (1970)

71. MULHOLLAND, H. and JONES, C. R., *Fundamentals of Statistics*, Butterworths, London (1968)

72. NEAME, K. D. and HOMEWOOD, C. A., 'Rapid Determination of Balance Point in Multichannel Liquid Scintillation Counters', *Anal. Biochem.*, **49**, 511–516 (1972)

72a. NEAME, K. D. and HOMEWOOD, C. A., 'Inexpensive Liquid Scintillation Counting of Aqueous Samples', *Anal. Biochem.*, **57**, 623–627 (1974)

73. NEWBERY, G. R., *Standards of Activity*, Review 4, revised edn, The Radiochemical Centre, Amersham, England (1968)

74. NOUJAIM, A., EDISS, C. and WIEBE, L., 'Precision of Some Quench Correction Methods in Liquid Scintillation Counting', *Organic Scintillators and Liquid Scintillation Counting*, 705–717, Academic Press, New York (1971)

75. OBER, R. E., HANSEN, A. R., MOURER, D., BAUKEMA, J. and GWYNN, G. W., 'Oxygen Flask Combustion Analysis of ^{14}C and ^{3}H: Description of Method and Apparatus and Summary of Experience, with Emphasis on Analysis of Low Levels of Activity', *Int. J. appl. Radiat. Isotopes*, **20**, 703–709 (1969)

76. PATTERSON, M. S. and GREENE, R. C., 'Measurement of Low Energy Beta-emitters in Aqueous Solution by Liquid Scintillation Counting of Emulsions', *Anal. Chem.*, **37**, 854–857 (1965)

77. PAUS, P. N. 'Solubilization of Polyacrylamide Gels for Liquid Scintillation Counting', *Anal. Biochem.*, **42**, 372–376 (1971)

78. PENG, C. T., 'A Review of Methods of Quench Correction in Liquid

Scintillation Counting', *The Current Status of Liquid Scintillation Counting* (E. D. Bransome, Ed), 283–292, Grune and Stratton, New York (1970)

79. POLLAY, M. and STEVENS, F. A., 'Solubilization of Animal Tissues for Liquid Scintillation Counting', *The Current Status of Liquid Scintillation Counting* (E. D. Bransome, Ed), 207–211, Grune and Stratton, New York (1970)

80*. PRICE, L. W., 'Practical Course in Liquid Scintillation Counting. Part I. Principles and Chemistry', *Laboratory Practice*, 22, 27–31 (1973)

81*. PRICE, L. W., 'Practical Course in Liquid Scintillation Counting. Part II. Preparing of Samples – 1', *Laboratory Practice*, 22, 110–114 (1973)

82*. PRICE, L. W., 'Practical Course in Liquid Scintillation Counting. Part III. Preparation of Samples –2', *Laboratory Practice*, 22, 181–194 (1973)

83*. PRICE, L. W., 'Practical Course in Liquid Scintillation Counting. Part IV. The Practical Counter and Quench Correction', *Laboratory Practice*, 22, 277–283 & 296 (1973)

84*. PRICE, L. W., 'Practical Course in Liquid Scintillation Counting. Part V. Computing and Calculating Techniques', *Laboratory Practice*, 22, 353–358 (1973)

84a*. PRICE, L. W., 'Practical Course in Liquid Scintillation Counting. Part VI. Choosing and using Liquid Scintillation Counting Equipment – 1', *Laboratory Practice*, 22, 417–422 (1973)

84b*. PRICE, L. W., 'Practical Course in Liquid Scintillation Counting. Part VII. Choosing and using Liquid Scintillation Counting Equipment – 2', *Laboratory Practice*, 22, 480–484 (1973)

84c*. PRICE, L. W., 'Practical Course in Liquid Scintillation Counting. Part VIII. Hazard and Safety Aspects of Liquid Scintillation Counting', *Laboratory Practice*, 22, 571–574 (1973)

85. RADIN, N. S., 'Round-table on "Methods of Counting Acids and Other Substances by Liquid Scintillation" ', *Liquid Scintillation Counting* (C. G. Bell and F. N. Hayes, Eds), Pergamon Press, Oxford (1958)

86. RAPKIN, E., *Temperature Control in Liquid Scintillation Counting*, Digitechnique Review, Intertechnique, 78-Plaisir-France (1968)

87*. RAPKIN, E., *Sample Preparation for Liquid Scintillation Counting. No. 1. Solubilization Techniques*, Digitechnique Review, Intertechnique, 78-Plaisir-France (1969)

88*. RAPKIN, E., *Sample Preparation for Liquid Scintillation Counting. Part 2. Solvents and Scintillators*, Digitechnique Review, Intertechnique, 78-Plaisir-France (1970)

89. *Recommendations of the International Commission on Radiological*

Protection, I.C.R.P. Publication 2, Report of Committee II on Permissible Dose for Internal Radiation, 1959, Pergamon Press, Oxford (1960)

90. *Recommendations of the International Commission on Radiological Protection, I.C.R.P. Publication 3*, Report of Committee III on Protection against X—rays up to Energies of 3 MeV and Beta— and Gamma—rays from Sealed Sources, 1960, Pergamon Press, Oxford (1960)

91. *Recommendations of the International Commission on Radiological Protection, I.C.R.P. Publication 6* (as amended 1959 and revised 1962), Pergamon Press, Oxford (1964)

91a. *Recommendations of the International Commission on Radiological Protection, I.C.R.P. Publication 9*, Pergamon Press, Oxford (1966)

92*. REES, D. J., *Health Physics*, Butterworths, London (1967)

93. ROGERS, A. W. and MORAN, J. F., 'Evaluation of Quench Correction in Liquid Scintillation Counting by Internal, Automatic External, and Channels' Ratio Standardization Methods', *Anal. Biochem.*, **16**, 206–219 (1966)

94*. SCHRAM, E., *Organic Scintillation Detectors*, Elsevier, Amsterdam (1963)

95. SHAW, W. A. and HARLAN, W. R., 'An Improved Method for Counting Radioisotopes Absorbed on Silica Gel', *Anal. Biochem.*, **43**, 119–122 (1971)

96. SHNEOUR, E. A., ARONOFF, S. and KIRK, M. R., 'Liquid Scintillation Counting of Solutions containing Carotenoids and Chlorophylls', *Int. J. appl. Radiat. Isotopes*, **13**, 623–627 (1962)

97. TAKAHASHI, H., HATTORI, T. and MARUO, B., 'Liquid Scintillation Counting of C^{14} Paper Chromatograms', *Anal. Biochem.*, **2**, 447–462 (1961)

98. TISHLER, P. V. and EPSTEIN, C. J., 'A Convenient Method of preparing Polyacrylamide Gels for Liquid Scintillation Spectrometry', *Anal. Biochem.*, **22**, 89–98 (1968)

99. TURNER, J. C., *Sample Preparation for Liquid Scintillation Counting*, Review 6, The Radiochemical Centre, Amersham, England (1967)

100*. TURNER, J. C., *Sample Preparation for Liquid Scintillation Counting*, Review 6 (revised edn), The Radiochemical Centre, Amersham, England (1971)

101. *U.S. Congressional Hearings on Nature of Radiation Fallout and Its Effects on Man* (1956); Report of the United Nations Scientific Committee on the Effects of Atomic Radiation, New York, 1958. Quoted by Setlow, R. B. and Pollard, E. C., *Molecular Biophysics*, Addison-Wesley, Reading, U.S.A. (1962)

102*. VENVERLOO, L. A. J., *Practical Measuring Techniques for Beta Radiation*, Macmillan, London (1971)

103. WALTER, W. M. and PURCELL, A. E., 'Elimination of Color Quench in Liquid Scintillation Counting of ^{14}C–Carotenoids', *Anal. Biochem.*, **16**, 466–473 (1966)

104*. WANG, C. H. and WILLIS, D. L., *Radiotracer Methodology in Biological Science*, Prentice-Hall, New Jersey (1965)

105. WATERFIELD, W. R., SPANNER, J. A. and STANFORD, F. G., 'Tritium Exchange from Compounds in Dilute Aqueous Solutions', *Nature (Lond.)*, **218**, 472–473 (1968)

106. WHISMAN, M. L., ECCLESTON, B. H. and ARMSTRONG, F. E., 'Liquid Scintillation Counting of Tritiated Organic Compounds', *Anal. Chem.*, **32**, 484–486 (1960)

107. WHITE, D. R., 'An Assessment of the Efficiencies and Costs of Liquid Scintillation Mixes for Aqueous Tritium Samples', *Int. J. appl. Radiat. Isotopes*, **19**, 49–61 (1968)

108*. WILSON, B. J. (Ed), *The Radiochemical Manual*, 2nd edn, The Radiochemical Centre, Amersham, England (1966)

109. WILSON, S. H. and KRONICK, M. N., 'New Assay Technique for Reactions that Produce Radioactive Gases', *Anal. Biochem.*, **43**, 460–467 (1971)

110. WINKELMAN, J. and SLATER, G., 'Chemiluminescence of Liquid Scintillation Mixture Components', *Anal. Biochem.*, **20**, 365–368 (1967)

111. WYLD, G. E. A., 'Statistical Confidence in Liquid Scintillation Counting', *The Current Status of Liquid Scintillation Counting*, (E. D. Bransome, Ed), 69–75, Grune and Stratton, New York (1970)

112 YAKUSHIN, F. S., 'Application of the Kinetic Isotope Effect of Tritium to the Investigation of the Mechanism of Hydrogen Substitution and Transfer Reactions', *Russ. Chem. Rev.*, **31**, 123–131 (1962)

113. ZELDIN, M. H. and WARD, J. M., 'Acrylamide Electrophoresis and Protein Pattern During Morphogenesis in a Slime Mould', *Nature (Lond.)*, **198**, 389–390 (1963)

The following are some books on the general use of radioisotopes in chemical investigations: *see* also Ref. 104 above.

Carbon–14 Compounds, J. R. CATCH, Butterworths, London (1961)
Chemical Applications of Radioisotopes, H. J. M. BOWEN, Methuen, London (1969)

Isotopic Carbon, M. CALVIN, C. HEIDELBERGER, J. C. REID, B. M. TOLBERT, and P. E. YANKWICH, John Wiley, New York (1949)

Isotopic Tracers in Biochemistry and Physiology, J. SACKS, McGraw-Hill, New York (1953)

Principles of Radiochemistry, H. A. C. McKAY, Butterworths, London (1971)

Principles of Radioisotope Methodology, 3rd edn, G. D. CHASE and J. L. RABINOWITZ, Burgess, Minneapoli (1967)

Radioisotope Laboratory Techniques, 3rd edn, R. A. FAIRES and B. H. PARKES, Butterworths, London (1973)

Radioisotope Techniques, R. T. OVERMAN and H. M. CLARK, McGraw-Hill, New York (1960)

Techniques of Autoradiography, A. W. ROGERS, Elsevier, Amsterdam (1967)

Techniques of Radiochemistry, S. ARONOFF, Iowa State College Press, Ames (1956)

Tracer Applications for the Study of Organic Reactions, J. G. BURR, Interscience, New York (1957)

Tritium and its Compounds, E. A. EVANS, Butterworths, London (1966)

Index

Page numbers in *italic type* indicate tables or illustrations